SCIENCE

It's Changing Your World

Easy does it! A robotic hand picks up an egg—very gently. Robots, important in industry, may soon serve in your home.

HANK MORGAN/RAINBOW

□ BOOKS FOR WORLD EXPLORERS
□ NATIONAL GEOGRAPHIC SOCIETY

CONTENTS

Copyright © 1985 National Geographic Society
Library of Congress CIP Data: page 104

In a special chamber, gas and electricity carve tiny electrical circuits into silicon wafers. Each of the 13 wafers measures 3$^1/_2$ inches (9 cm)* across. The wafers replace wires and silicon chips. A single wafer holds all the circuitry used in the most powerful computers.

*Metric figures in this book have been rounded off.

COMPUTERS *and* LASERS

Tools of the Space Age

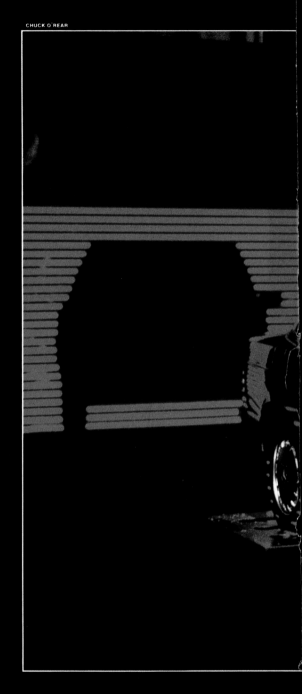

CHUCK O'REAR

A huge explosion is taking place at this very moment. You can't hear it. Still, it's more powerful than any other explosion in history. It's an explosion of knowledge.

Scientists give this comparison: If you are 11 years old, more knowledge has been gained in your lifetime than in *all the previous history of the earth.* What's more, scientists say, in the next ten years or so, our knowledge of the universe will probably double again. They estimate that of all the scientists who ever existed, nearly 90 percent are alive and at work today in laboratories, classrooms, and offices.

Two tools have contributed greatly to the knowledge explosion. They have helped develop the knowledge, and they have helped put it to use. The tools are the computer and the laser. They have countless uses. Yet they hardly existed when your parents were your age.

Many scientists call the computer the most important development of this century. This machine has come a long way since it first appeared. The earliest computer, built in the 1940s, weighed 30 tons (27 t) and filled a large room. A personal computer today can do more jobs than that giant. Yet some are smaller than a big-city telephone book. Hand-size computers are on the way.

What exactly do computers *do?* They store huge amounts of information and process it in a very short time. They shuffle the information and sift through it. They match it and compare it. They analyze it and organize it. They provide the computer operator with the best answer—or selection of answers—to a problem.

Of course, computers would be useless without people to program them. Computers are not "smart" by themselves. They can simply

4

Using laser light, an engineer in the Federal Republic of Germany makes measurements of a scale-model Mercedes-Benz. The measurements are extremely accurate. The engineer will use the results to make design changes aimed at improving the performance of the full-size car.

Love at first byte? A young visitor to the Capitol Children's Museum, in Washington, D. C., meets a computer (left). At the museum, children play with computers and begin learning about the hundreds of jobs they can do.

At Stanford University, in Stanford, California, kids teach teachers how to teach computer skills to kids (big picture). Students from Jordan Middle School, in Palo Alto, do the instructing in the summer. In the small picture, Kristin Chandler helps Betty Allen master a computer command.

process more information more rapidly than people can—and without forgetting anything. A computer cannot solve a problem if someone doesn't tell it what needs solving in the first place. And computers would not even exist but for the human brain, the most complex thing in the known universe.

Like the computer, the laser has become a vital tool in today's world. It was once viewed as a toy, something of no real use. Now, laser beams do important jobs in science, industry, and medicine.

Laser light cuts, mends, guides, transmits, measures, and vaporizes. Depending on its strength, a laser beam can cut through steel or mend delicate eye parts. Laser beams measure the distances of faraway planets to within millimeters. They clear

Working with a computer, Tei Gordon, 13, helps big companies save millions of dollars in heating and cooling costs. He operates a weather center from his home in Corvallis, Oregon. Tei feeds into the computer the official temperature readings for 210 United States cities. The machine sorts the readings according to Tei's commands. The results help clients discover if their buildings are using too much energy for heating and cooling. Here, Tei reads a printout of his weekly report. He makes about $2,000 a year with his weather service.

particles of dirt from valuable oil paintings. A single beam can transmit a million kilowatts of laser power in a pulse shorter than a billionth of a second. Two hair-thin beams can carry more than 8,000 telephone calls at one time—and the number will soon jump to 24,192.

Using their knowledge and imagination—and aided by such machines as computers and lasers—scientists are developing new ways to build things . . . to produce energy . . . to improve means of travel and communication . . . to fight disease and overcome disabilities . . . to learn more about the far reaches of space. Almost too fast to imagine, science is changing your world. Want to keep pace? Read on. . . .

1

INDUSTRY
Building Tomorrow Today

Not long ago, robots belonged only to the world of science fiction. Today, they do big jobs in many industries. Every day, you probably use objects that robots helped build. Here, Kenneth Overton, of the University of Massachusetts robotics lab, in Amherst, works a robotic hand. He's developing fingertip sensors that will enable a robot to identify an object by its size and shape. As technology advances, robots and other space-age tools will become more and more important in industry and in people's everyday lives.

9

Getting ready to serve, a steel-and-plastic waiter (left) takes on an order at the Two Panda Robot Restaurant, in Pasadena, California. The restaurant owners bought two robots, Tanbo R-1 and Tanbo R-2, for less than it costs to pay two people to work for a year. "They never miss work," says co-owner Shayne Hayashi, shown loading the robot's tray. Human waiters take orders from customers. The robots deliver the food. They also sing and tell jokes in five languages.

New teacher: Second graders Joy Yamamura and Jennifer Kitson, both 8, meet a robot named Topo (right). Controlled by a computer, Topo rolls on wheels and talks. The robot was designed to introduce youngsters to the machines that are in their future. Joy and Jennifer attend Stevens Creek Elementary School, in Cupertino, California.

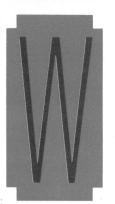

When diners in Pasadena, California, enter the Two Panda Robot Restaurant (above) for the first time, they're in for an unusual experience. They see two robots whirring along the aisles, carrying orders to customers.

"Hello," say the robots as they deliver orders of cheeseburgers or egg rolls. "How are you?" The robots may burst into song or tell a joke. They speak English, French, German, Spanish, and Chinese. Their hearts are computers. Programmed by humans, the computers direct the robots' activities.

"Tanbo R-1 and Tanbo R-2 make a hit with customers," says Shayne Hayashi, co-owner of the restaurant. "We still use human waiters to take orders and do other work. But the robots serve and entertain.

They don't ask a salary. They need only an overnight battery charge to keep them running. In just a short while, they've more than paid for themselves."

The scene at the Two Panda may be unusual today. But it will become more and more common in years to come. In the home, robots may do the dusting and vacuuming. They may wake you up in the morning and serve you breakfast in bed. In shops, offices, factories, and fields, robots will do many jobs that people find boring, difficult, or dangerous. Because the jobs *are* of that nature, robots often do them better than humans. Robots have no minds to wander or worry. They always do exactly what they're told. In fact, that's all they *can* do.

Tens of thousands of robots are now at work in many industries. The robots aren't humanoid—that is, they don't much resemble people. Industrial robots are usually headless box-shaped machines

10

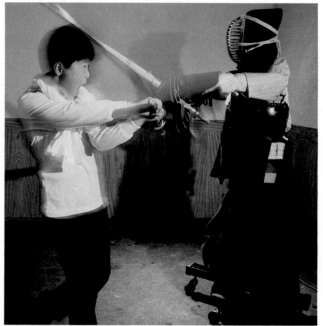

In Tokyo, a robot teaches kendo (above). Kendo—fencing with bamboo swords—is a Japanese martial art. The robot repeats a series of standard moves. Once the student understands the pattern, he can defeat the robot.

When OPD2 talks, kids listen (above). The robot works for the police department of Orlando, Florida. Here, OPD2 gives the word on bicycle safety. Behind the scenes, Officer Joseph Bongiorno works OPD2 by remote control. Bongiorno started the robot program.

Sparks fly as robots weld automobile bodies at the General Motors plant in Lordstown, Ohio (right). Robots also paint and assemble parts and lift and move heavy loads. They do jobs that people find boring, difficult, or dangerous.

with one or two arms attached. But, like humans, they work hard—and they do hard work.

Take the Chrysler automobile plant in Windsor, Ontario, in Canada. Plant manager George H. Hohendorf calls it "one of the most advanced high-technology assembly operations in North America." At the plant, 129 robotic arms dart about the cars as they move down the assembly line. The robots weld a car body together in about a minute. They paint. They apply protective coatings. They lift assembled parts and place them on overhead transporters for further work by other robots.

"The robots produce work of high quality," says Hohendorf. "And they're highly dependable. They do

their jobs right on time to a firm production schedule. Everything goes like clockwork. We can tell our suppliers far in advance exactly when to deliver materials. There is no waiting, no delay. That pushes costs down all around."

At the General Electric dishwasher factory in Louisville, Kentucky, a master computer keeps tabs on the work of 229 robots. They mold the plastic tubs. They paint dishwasher parts. They assemble metal panels. They lift the dishwashers on and off conveyors. Then they package the appliances.

"The robots do their jobs so well that repairs on dishwashers have been cut in half," says GE official Ray Rissler. "Robotized factories can quickly make small changes in a product. Even big changes—switching from one product model to another, for example— can be made with a few computer commands."

No matter what a robot does or what it looks like, it nearly always has a computer for a heart. But here's a twist: At Apple Computer, Inc., in Fremont, California, robots *build* computers.

One robot assembles nearly all the 100 parts of the Apple's logic board. That's the part of the computer that does the "thinking." Another robot plugs 32,000 electronic parts into circuit boards every hour. Other robots assemble the boards and other parts into working computers.

Robots had a part in making the 13 million computers now in American households. Robots will help build the 62 million more computers expected to be

From a central computer, a technician supervises production of dishwashers at the General Electric factory in Louisville, Kentucky. His workers are robots—hundreds of robots. Only a few humans monitor them. The robots cost about $100,000 apiece. They quickly pay for themselves, plant officials say.

in homes by 1995. "That figure is important," says analyst Hilda Uribe, of Future Computing, a research group in Dallas, Texas. "It means that by then almost as many homes will have a computer as have a telephone." Already 75 percent of the nation's high schools have computers for their students. Some colleges require each student to have a computer.

Soon, robots will do more than work on assembly lines. "I foresee robots programmed to pick oranges, to pump gas, even to assist in brain surgery," says robotics pioneer Joseph Engelberger. He founded Unimation, Inc., in Danbury, Connecticut. It's the world's largest manufacturer of industrial robots.

"By the year 2000," Engelberger predicts, "robots will be performing nearly all manufacturing jobs. They'll be tending crops and livestock. And they'll be gathering needed metals on the ocean floor and deep inside the earth. They'll see, talk, hear, and plan, and they'll learn from experience."

What about people? Will there be jobs for them if such advanced robots are developed? Some experts fear that robots will put humans out of work.

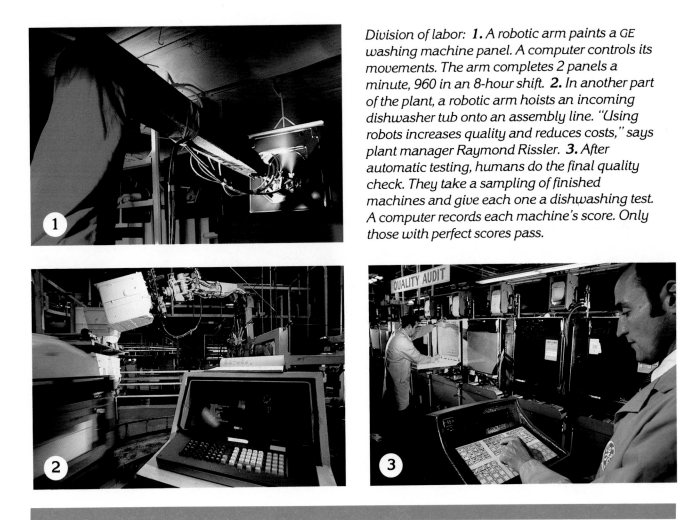

Division of labor: **1.** *A robotic arm paints a GE washing machine panel. A computer controls its movements. The arm completes 2 panels a minute, 960 in an 8-hour shift.* **2.** *In another part of the plant, a robotic arm hoists an incoming dishwasher tub onto an assembly line. "Using robots increases quality and reduces costs," says plant manager Raymond Rissler.* **3.** *After automatic testing, humans do the final quality check. They take a sampling of finished machines and give each one a dishwashing test. A computer records each machine's score. Only those with perfect scores pass.*

Engelberger believes not. "Robots actually have created more jobs than they have taken away," he says. "Engineers are needed to design robots. Technicians are needed to put them together. Computer programmers and other experts are required to make them work. That will continue to be the case. The types of jobs people do will change. The demand for workers will not."

Two jobs—cutting and drilling—are a part of most industries' work. Automobile factories, for example, must have cutters to trim unwanted metal from molded parts. Drills must bore hundreds of holes into car parts for assembly. But cutters and drills become dull. They overheat. They break.

Now many industries are using lasers to do these jobs. Quickly but delicately, laser beams can strip the husks off peanuts. They can also slice through steel. Baby-bottle makers use lasers to drill holes in nipples. Clothing manufacturers use them to cut fabric. The beams slice cleanly through dozens of layers of cloth.

Laser beams have other uses as well. In many

The robot will be easily programmable through a keyboard or with recorded programs. With its computer intelligence, it will reason and plan. It will adjust programs to fit new circumstances.

TV-camera laser eyes will give the robot a 3-D view of its surroundings. A TV monitor will show what the eyes take in.

Sensitive fingertips and flexible joints will enable the robot to do delicate work. It will pick up, identify, sort, and use all kinds of objects without dropping them or crushing them.

YES, I WASH WINDOWS TOO

The robot will hear, answer, and respond to spoken commands. It will have a vocabulary of 10,000 or more words and will translate foreign languages. It will even learn to recognize different accents.

An easy-to-use remote-control system will enable all the family to activate the robot's programs.

JOHN HUEHNERGARTH

Tomorrow's household robot

Will a robot butler someday serve your dinner? And will a robot maid then wash the dishes? Robotics engineers think it's likely. They're working to make robots that can do an even wider range of jobs than they do today. Robots of the future will have all the five senses. They will be equipped with highly advanced computers. They will be able to do a limited amount of reasoning and to learn from experience. Here, an artist's cartoon shows some of the features you'll find on a robot of the future.

16

Sensitive project: Scientists at the Massachusetts Institute of Technology, in Cambridge, work to give robots artificial senses. John Purbrick (above) holds a robotic hand with sensors attached. Hundreds of bumps dot the sensors. A scanner reads the amount of pressure on the bumps when the sensors touch an object. Thin rubber sensors held by John Hollerbach (above) contain metal particles. When the sensors touch an object, the particles come together. Electricity flows through them. The flow helps make a computer *picture of the object. Tomás Lozano-Pérez (above) is writing a computer program that will allow a robot to recognize objects and avoid bumping into them. It's complicated work. But such a program will be an essential feature of many kinds of robots.*

areas, telephone signals are carried by laser beams rather than by electrical currents. The light speeds cross-country through flexible glass threads.

The threads are called optical fibers. Light travels through them the way water travels through a hose. Just two strands of optical fiber can carry 8,064 telephone conversations at the same time. Soon that number will leap to 24,192. You can see such fibers on page 20. Two copper wires the same size can carry only about two dozen conversations. Think about it the next time you pick up a telephone: Your words may be traveling by laser light!

Lasers also are used to make extremely accurate measurements. Chances are you'll use lasers this way in the career you choose. If you become an astronomer, you'll use them to tell you how far away a distant planet is. As a biologist, you'll use lasers to measure various microscopic creatures. As a space engineer, you'll use lasers to make sure all the parts of a space vehicle fit together exactly right.

With lasers, technicians can make a special type of three-dimensional image. It's called a hologram (HOLE-uh-gram). Holograms are produced on film, or on glass that has been specially treated. You need no special glasses or light to view a hologram. You may already have seen one of these 3-D images. A

17

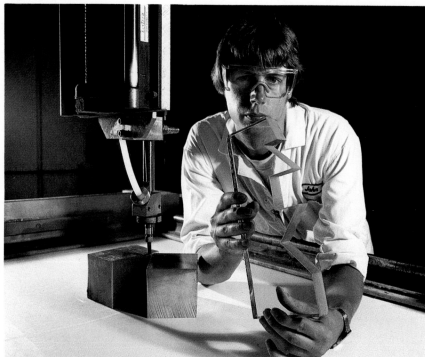

"Like cutting butter with a hot knife." That's how lab technician John Wilson describes working with the Waternife. This new tool uses a thin jet of high-pressure, gritty water to cut through aluminum and glass (above). Water without the grit slices pizza (above, right). "The Waternife cuts fast and clean and produces no heat," says Wilson, who works with Flow Systems, Inc., in Kent, Washington. "It's going to revolutionize a lot of industrial cutting."

Zapping dirt and old varnish, laser light cleans a painting (right). David Rayner, a chemist with Canada's National Research Council, in Ottawa, Ontario, observes the process. The beam apparently does no harm to the paint. To make the beam visible, the photographer moved a card at intervals along its path.

hologram of a bald eagle appeared on the cover of the March 1984 NATIONAL GEOGRAPHIC.

Holograms are fun to look at. Museums sometimes use them instead of displaying items that are very fragile or valuable. The images look real enough to touch. Holograms have other practical uses as well.

"At first, they were little more than items for amusement," says Karl Stetson, a senior research scientist with United Technologies Corporation, in East Hartford, Connecticut. "But then scientists and engineers found that they could put holograms to good use."

At United Technologies, engineers use holograms to detect faults inside jet engines. "The 3-D picture shows every detail from every angle," says Dr. Stetson. "It's like standing inside the engine for a look. Holograms let us see hidden problems."

Holograms may someday appear on paper money to foil, or defeat, counterfeiters. Holographic images already appear on some credit cards. "Engineers are finding new uses for holography all the time," says

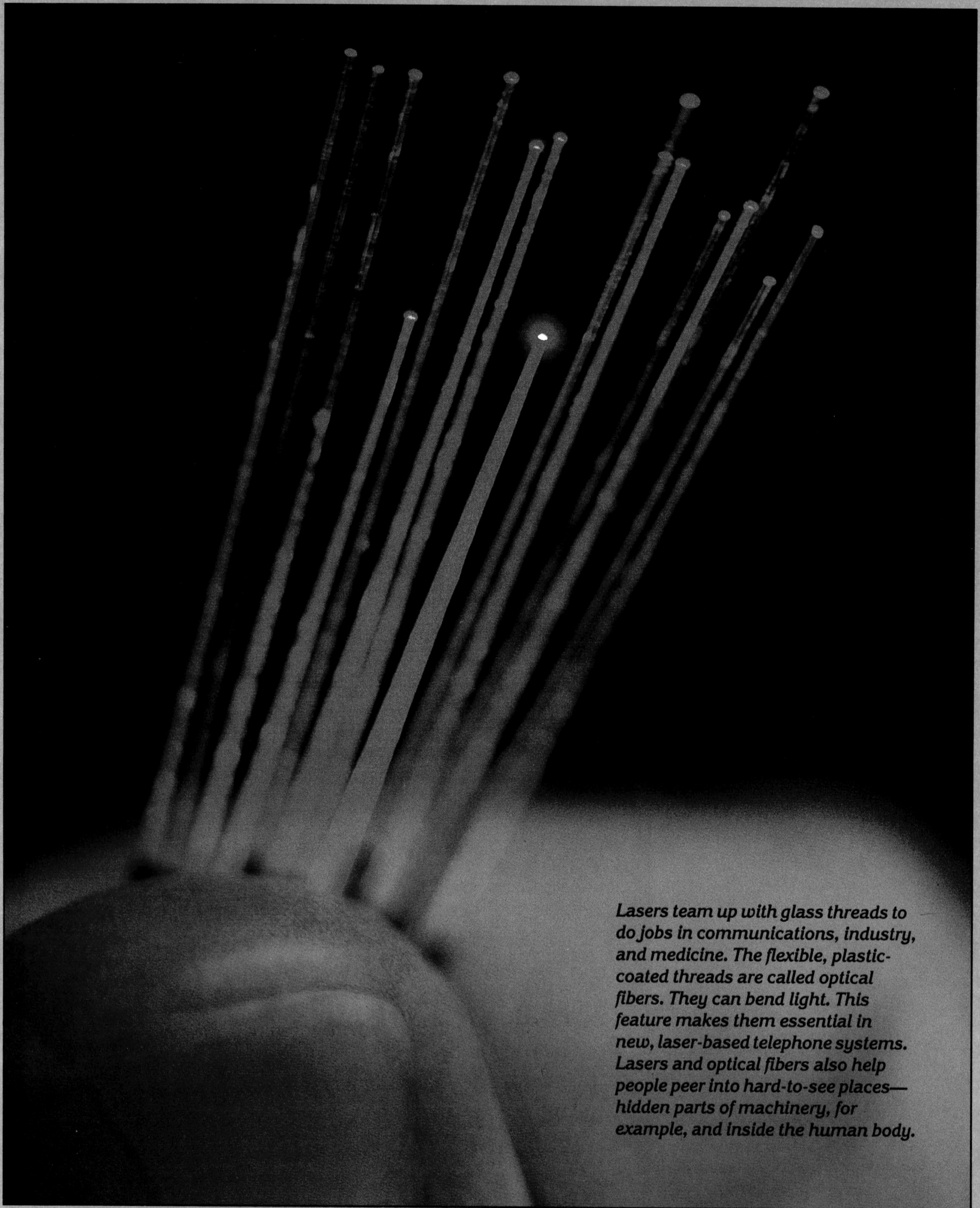

Lasers team up with glass threads to do jobs in communications, industry, and medicine. The flexible, plastic-coated threads are called optical fibers. They can bend light. This feature makes them essential in new, laser-based telephone systems. Lasers and optical fibers also help people peer into hard-to-see places—hidden parts of machinery, for example, and inside the human body.

CHARLES WEST/THE STOCK MARKET

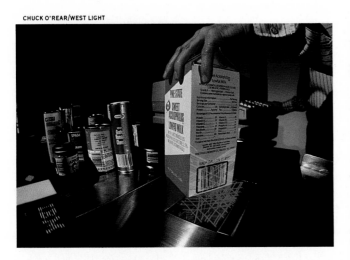

At a supermarket checkout, a laser reads a computer-coded price tag (above). When the first working laser flashed on in 1960, people didn't know quite what to use it for. Today, it helps people weld, measure, communicate, cut, vaporize—and, of course, throw light on things.

Mounted on a rail truck, lasers make exact measurements of a railroad tunnel. The information will tell railroad officials how high and wide they can load their freight cars.

Dr. Stetson. "We never know what's coming up next."

Realistic images of another kind help people learn to operate complicated and expensive equipment. Computers make these images, which are based on real scenes. The images are displayed in devices called simulators.

Suppose you go into astronaut training. It would cost many millions of dollars to send you into space on a training flight. It would be dangerous, too. Instead, you use a space-vehicle simulator at your training center. The simulator surrounds you with the sights, sounds, and motions of an actual flight into space.

"We program 6,800 malfunctions into the space simulator," says William Turner. He's president of Link Flight Simulator Division of The Singer Company, in Binghamton, New York. Link builds simulators for the National Aeronautics and Space Administration (NASA). "In a simulator," Turner continues, "everything goes wrong—on purpose. Astronauts learn how to deal with every imaginable emergency. Simulators help make actual launchings and space flight safe."

What other space-age products is modern industry

21

Scientists use lasers to create three-dimensional images for business and for pleasure. The images are called holograms. Printed on credit cards (left), holograms discourage attempts at counterfeiting. Above, holographer Fred Unterseher views a creation of his at the Museum of Holography, in New York City. Scott Lloyd, a museum official, describes holography's effect: "It's like taking away an object and leaving behind the light it reflects. Although the object is not there, you still can see it."

giving us? One Japanese company produces a talking camera. It tells you if you're about to make a bad picture. A humanlike voice warns you when, for example, there isn't enough light. Then the voice (produced by a computer) tells you to use the flash.

If a gabby camera isn't for you, how about a phone you can talk to? You can program one new model to recognize your voice. The phone remembers up to a hundred numbers. It dials them on command. Just pick up the receiver and say the name of the person you want to call. The phone takes it from there.

Space-age plastics make possible other amazing products. The U. S. Department of Agriculture, for example, is developing plastic crystals that have a special quality. They store or release heat as the weather changes. Fabric treated with the crystals can be worn in any climate. The same treated safari suit will keep you cool in a tropical jungle and warm on a freezing mountain peak.

Tougher, more heat-resistant plastics come on the market every few months. "The days of glass and metal may be numbered," says Laszlo Bonis, president of Composite Container Corporation, in Medford, Massachusetts. "By the 1990s, you'll find little of either in food packaging." Automobiles—including engines—also are changing over to plastic, part by part.

Plastics companies, electronics companies—businesses of all kinds are working hard to develop

"Take that, Darth Vader!" Cruz Fino, Jr., 14, pretends to be a spaceship commander as he shows how an arcade game of the future might look. A player would sit inside a mini-dome. The player would shoot laser beams at video targets projected onto the wall. Though bright, the beams would of course be of low power and harmless. Cruz, whose father is a technician with Laser Images, Inc., lives in Van Nuys, California.

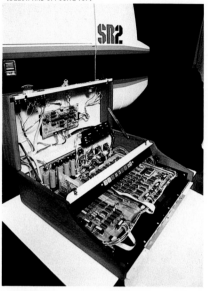

Shoot white-water rapids, go deep-sea diving, pilot a bobsled— all without leaving the theater. The SR2, a van-size theater (right, top), provides 12 wild rides and adventures. A computer (left) activates wind machines, movie equipment, and hydraulic lifts. They simulate, or copy, real action. At right, youngsters in the theater experience all the thrills of a roller coaster ride. Simulators are used for more than fun. They help teach the skills that drivers, pilots, and astronauts will need.

new and better products. In their operations, however, many factories create an unfortunate by-product. It's chemical wastes that find their way into the air, onto fields, and into streams. Such pollution can harm people. It can kill fish and other wildlife.

Scientists are hard at work finding ways to solve the pollution problem. One solution is especially exciting. It's the use of microscopic organisms called microbes. Microbes are all around us: in the water, in the soil, in the air. Some microbes love wastes, even toxic, or poisonous, wastes. They eat wastes and thrive on them.

"We could recruit this hidden work force," says microbiologist Brian J. Ford. Dr. Ford lectures at Cardiff University, in Cardiff, Wales, in Britain. "Microbes have always been nature's own pollution fighters," he continues. "Given encouragement, they could, in the future, clean up the environment for us."

One way to achieve that would be to reproduce great numbers of microbes in the laboratory, then inject the microbes into waste heaps. Another way would be to harness the power of naturally occurring

24

Electronic pollution

Machines, machines, machines—the modern world is full of them. They work for you. They bring you news and entertainment. They help keep you healthy. They also may give off radio and other waves that make other devices go haywire. The disturbance is called static or EMI— electromagnetic interference. Electric motors produce static. So do electronic devices—which are also disturbed by it. Affected equipment appears here in orange. How many static makers can you find? EMI can be merely annoying, as when it interferes with TV reception, or it can be serious. It may interfere with medical instruments or with an airplane's navigation equipment. Electronic pollution has become so troublesome that it has given rise to a new industry: the design and manufacture of static shields.

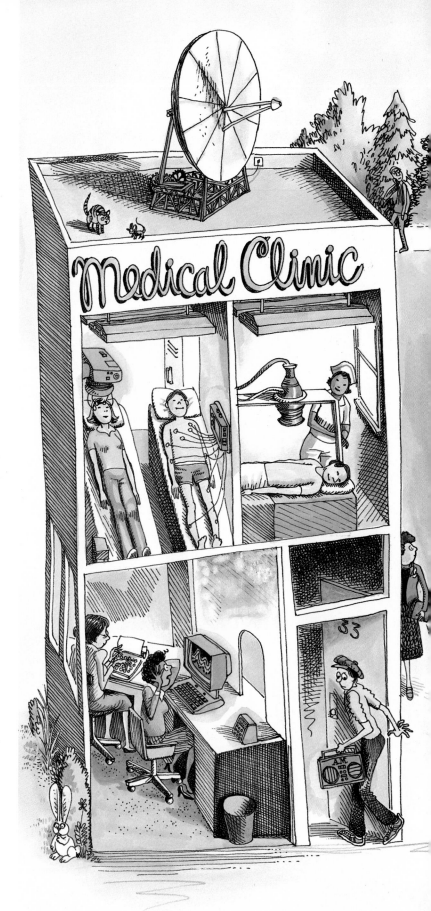

"super microbes." These microbes eat oil spills and turn poisons into useful substances.

"Toxic waste could actually become a valuable raw material," says Dr. Ford. "With microbes on our side, we'd be able to turn the waste into fertilizer that would help produce food for a growing population. Microbes could also help us convert waste cheaply into such fuels as alcohol—and even gasoline!"

Here, by studying the tiniest organisms, scientists are beginning to learn how to solve big problems. The knowledge they gain will become the basis for new technologies and new industries. The same is happening in many other fields. Knowledge opens doors. The doors lead to new worlds—and yet more doors.

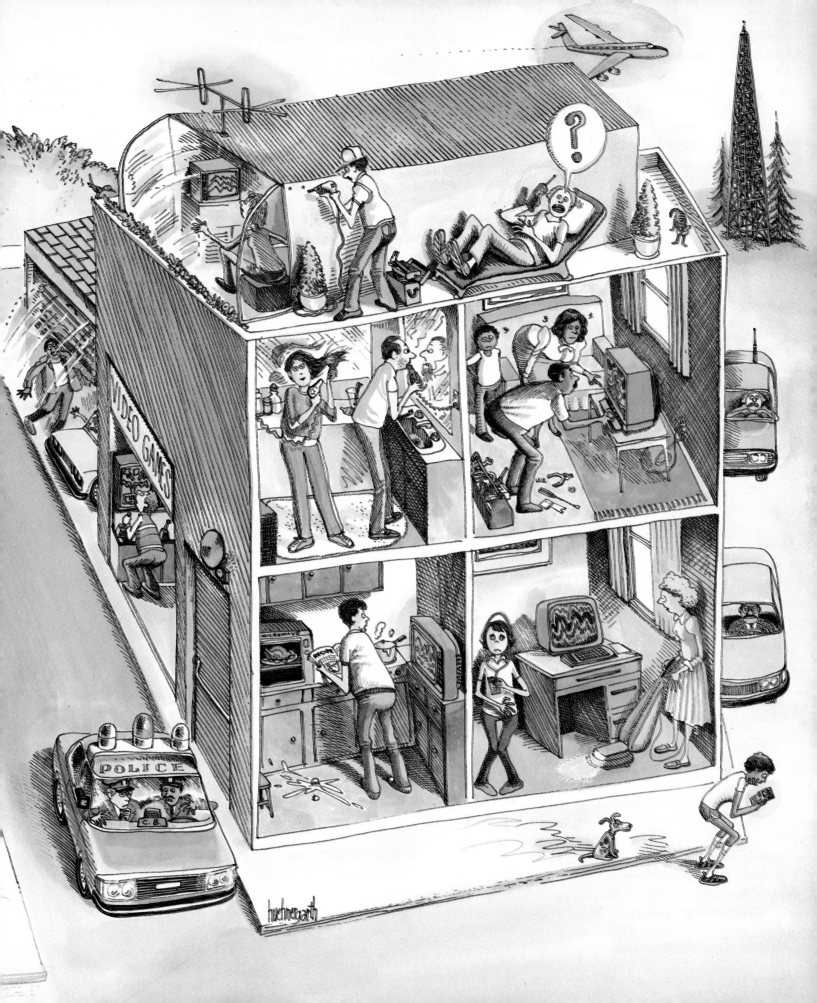

2

FUEL and FOOD

Energy for the Future

Machines and people need energy to keep going. As the world becomes more industrialized and as the population grows, the demand for energy increases. Scientists are seeking to develop cheaper, more efficient ways to produce fuel for machines and food for people. Here, in Altamont Pass, near San Francisco, California, an old force has been put to work. It's the wind. It powers windmills that generate electricity. In the coming years, the wind and other clean, renewable sources will fill more and more of our energy needs.

Powerful reflection: The sun provides the earth with its most plentiful source of energy. To make efficient use of that energy, scientists must catch the sun's rays, then direct and focus their energy. At Solar One, a power plant near Daggett, California, 1,818 truck-size mirror panels track the sun (above). They reflect its heat and light onto a boiler atop a tower. Resulting steam powers a machine called a turbine. It makes electricity. Heat piped to storage (right) powers the turbine at night.

Every living thing depends on it. It makes plants grow. It enables you to throw a baseball or figure out a math problem. It is energy—the ability to do work.

All the energy used by our bodies comes indirectly from the sun. Every day, the sun's rays flood the earth with energy. Plants absorb it and grow. Animals eat the plants, taking in their stored energy. Humans in turn get their energy from the meats and vegetables they eat.

The sun also produces most of the other kinds of energy we use. It heats air, causing the air to rise. That creates the winds that push boats and drive windmills.

It gives growth to trees used for firewood. By causing ocean water to evaporate, it brings about rain that collects as rivers. The rivers run huge electric plants. Fossil fuels—coal, oil, and natural gas—come from the remains of plants and animals that lived millions of years ago. The sun gave them life.

You could almost say that fossil fuels drive today's world. Americans, for example, get more than 80 percent of their energy from coal, oil, and natural gas. But the supply is running out. People are using fossil fuels much faster than nature is replacing them. Some experts estimate that a century from now the world will have run out of oil and natural gas. Coal will last longer, but someday it too will be gone.

When fossil fuels disappear, how will people make

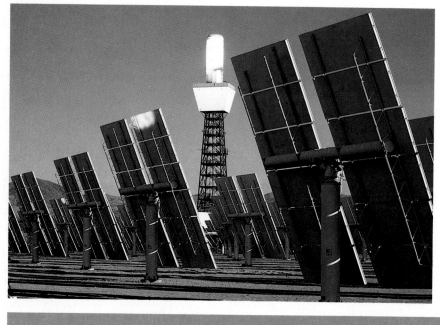

Controlled by a computer, Solar One's mirrors shift position slightly every two to ten seconds. That puts them constantly at just the right angle to reflect the sun's rays onto the boiler. So much focused light falls on the boiler that people can see its glow from 40 miles (55 km) away—far beyond the horizon. Solar One generates enough electricity to serve a town of 10,000 homes.

enough electricity to light their homes, to run computers, to operate factories? How will they fuel their automobiles?

"More and more, people will be using solar energy," says Frederick Morse. He directs the Office of Solar Heat Technologies of the U. S. Department of Energy, in Washington, D. C. "People will use the heat, light, wind, and water power that come from the sun," Dr. Morse continues. "Solar energy is clean, and the supply won't run out for billions of years."

In less than an hour, enough solar energy reaches the earth to fill everyone's energy needs for a full year. But the energy is spread out over land and sea. The challenge lies in harnessing the energy and putting it to work. Scientists now are finding ways to do just that.

Simply gathering the sun's heat is one way. Many modern houses have large rooftop solar collectors. The more sunlight there is, the more heat they absorb. Water tanks store the heat. Pipes circulate it as needed. In winter, the solar heat can warm the houses day and night. It can provide hot water the year round.

Solar energy can also be converted directly to electricity. Special cells made of silicon do the converting. Silicon is the main substance of sand. After oxygen, it is the most plentiful element on earth. You may have a calculator that runs on small solar cells. Larger cells grouped in panels can do heavier jobs. Solar panels provide satellites with electric power. The panels have powered the flight of an airplane. You'll see that plane on page 57.

In 1982, a solar-powered car named *The Quiet Achiever* rolled across Australia, a trip of 2,566 miles (4,130 km). "We felt the sun's power pushing us along, the way you feel daylight's warmth on a frosty morning," says Hans Tholstrup, who got the idea for the *Achiever*. No speed demon, the car cruised at 15 miles an hour (25 km/h). Its top speed was 40 miles an hour (65 km/h). The car had to be lightweight and energy-efficient, so it was built with neither springs nor air-conditioning. It wasn't exactly tops in comfort.

But comfort wasn't the point of the trip. "We proved it was possible to travel great distances on the

32

Solar power goes to sea

Sun and sailing go together—in more ways than one. The sun's power creates the winds that can push a boat along. Now the sun can actually help operate equipment on the boat. Solar cells on the deck of this sailboat convert sunlight directly to electricity. The electricity charges a battery that powers a radiotelephone, navigation and cabin lights, and perhaps television and stereo equipment. The buoy on the boat's right marks shallow water. It is equipped with solar cells. Throughout the night, its electric light will burn brightly on energy collected during the day. Ashore, solar heat collectors provide home heating and keep hot-water supplies steaming. At sea and on land, in the wilderness and in the city, solar cells and collectors are providing electricity and storable heat for people. The equipment does so without turbines or generators, smoke or noise.

JOHN HUEHNERGARTH

Sign of the times: As cattle graze below it, an advertisement on an abandoned windmill tower announces an additional use for the land (left). The mill once pumped water. Now buyers will build electricity-generating windmills at this Altamont site. Strong winds blow here the year round.

A line of windmills stretches across a hilltop at Altamont. One mill can generate enough electricity to supply the needs of 30 homes. Oil burned to light and heat an equal number of homes dumps a ton of smoke and grime into the air in a year. Windmills produce no pollution.

sun's energy," says Tholstrup. "Now it's up to other people to prove it's practical."

Making solar energy practical for widespread use is a major aim of many energy companies. In California, two experimental solar power plants now produce electricity that helps light people's homes. One plant, in Hesperia, uses solar cells to convert sunlight directly to electricity. The other, in Daggett, converts solar heat to steam. The steam drives turbines—machines that generate electricity. "Solar power is clean, quiet, and nearly maintenance-free," says William Bicker, an official with ARCO Solar, Inc., which owns the Hesperia plant. "There's no muss, no fuss, no bother."

Electricity can also be generated by the wind. People learned long ago that the wind could pump water or grind grain by turning the blades of windmills. When fossil fuels came into use, however, windmills lost importance. Now they're making a comeback. In 1981, the world's first commercial wind farm began whirring out electricity near Goldendale, Washington. Now more than a hundred wind farms generate electricity in breezy parts of a dozen or more states.

Another form of energy comes from the mineral uranium. It is nuclear energy. Uranium is in a constant state of slow decay. Its atoms are fissioning, or splitting, a few at a time. The decay—radioactivity—produces heat. In a nuclear reactor, many atoms are made to split at the same time. The fission releases enormous amounts of energy from a small amount

of fuel. A pound of enriched uranium produces nearly 70,000 times the energy of a pound of coal.

The world's first commercial nuclear power plant went into operation in Shippingport, Pennsylvania, in 1957. Today, 83 working nuclear power plants are scattered around the country. They provide Americans with 13 percent of their electricity. With added plants, energy experts expect that figure to reach 20 percent by 1990. In some countries, such plants provide nearly half the total electric power.

Another nuclear fuel waits to be put into general use. It's a form of hydrogen called deuterium (dew-TEER-ee-um). The oceans hold vast—you could say limitless—quantities of it. "It's an inexpensive fuel, but an expensive and complicated device is needed to burn it," says Harold Furth. He's director of the

Princeton Plasma Physics Laboratory, in New Jersey.

To release deuterium's power, scientists must create temperatures of about 200 *million* degrees Fahrenheit. "That temperature changes deuterium into super-hot gas we call plasma," says Dr. Furth. In the plasma state, the nucleus, or core, of one atom merges with the nucleus of another. The merging, called fusion, produces vast amounts of heat. The heat can make steam for generating electricity.

"After 30 years of research, we are finally able to reach the temperatures required for fusion," says Dr. Furth. "We expect to find a way during the 1980s to get as much power *out* of the fuel as we must put in to make it hot." Commercial fusion reactors, he believes, may become practical early in the next century.

Radioactive materials *(Continued on page 41)*

Repairing a windmill's generator, a workman seems small against the mill's 28-foot (9-m) blades. This tower equals the height of a six-story building. Federal law requires power companies to buy electricity from anyone producing it with windmills. Now thousands of windmills whir out electricity in California. Hundreds more operate in many states from Vermont to Hawaii.

Giant eggbeaters? No, they're Darrieus rotors, but they are often called eggbeaters, and they do whip up a lot of energy. Darrieus rotors are among the most efficient windmills designed. After a motor starts it spinning, a rotor accepts wind from any direction. About 160 of these rotors stand at Altamont. The generators and control devices all rest on the ground, making repairs easier than with other types of windmills.

To squeeze the most power from the wind, standard windmills need constant monitoring. In the control room of U. S. Windpower, Inc., at Altamont (below), 27 computers record the performance of 1,336 windmills. In addition, a computer on each windmill tower keeps check on the wind and adjusts the direction and angle of the blades.

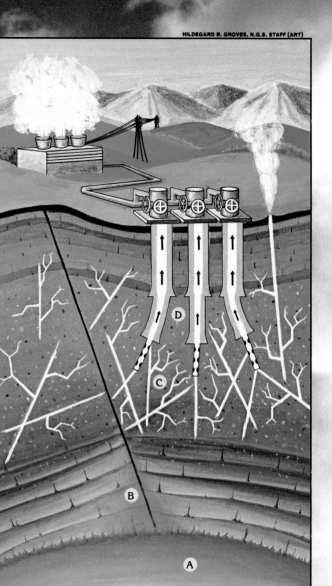

Steam from deep underground billows out of stacks at The Geysers, a power plant near Santa Rosa, California (big picture and far right). The Geysers uses the earth's own heat, called geothermal energy, to produce electricity. The heat comes from the center of the earth. It is caused by radioactivity, the slow decay of certain materials. The heat melts rock. In a few places, the melted rock, called magma (A, above), comes within a few miles of the surface. Layers of rock above it become very hot. Sometimes the rock shifts during earthquakes. That creates big cracks called faults (B), and many smaller fractures (C). Faults often extend to the surface; the fractures sometimes do. Surface water

runs into the faults and seeps into the fractures. It becomes heated. Pressure builds up. When drilling equipment bores through the rock and finds a fracture, hot water, and sometimes steam, shoots up the pipes (D) to the surface. At The Geysers, the piped steam drives turbines. The electricity produced serves a million people. In Japan, the Philippines, Iceland, Italy, Mexico, and New Zealand, as well as the United States, people use more and more geothermal energy every year. That eases the demand for oil and other fuels.

Running a fish farm in the desert may not sound like a promising business. But scientists at the Environmental Research Lab, in Tucson, Arizona, are showing how raising tilapia (above) can make profits. Several kinds of the fast-growing fish are cultivated in tanks inside greenhouses (left). The scientists develop the superior varieties. Unlike many kinds of

fish, tilapia (tih-LAHP-ee-uh) breed year-round. A female may produce a thousand fry, or young, every year. In six months, the fry reach 1 pound (1/2 kg)—harvesting weight. Tilapia's firm flesh (above) and delicate flavor please diners. The fish-tank water is recycled. Waste-rich, it irrigates and fertilizes desert crops.

(Continued from page 35) decaying at the earth's core create a furnace of melted rock. In some places, the molten rock, called magma, bulges to within a mile or so of the surface. Engineers drill holes into hard rock above these places to find steam or super-hot water. Naturally occurring high pressure forces the steam or water to the surface. There it spins turbines.

This energy is called geothermal—earth heat. The largest geothermal power plant in the world operates in California. Named The Geysers, it lies 75 miles (121 km) north of San Francisco. The plant today generates enough electricity for a million and a half people. By 1990, according to plant officials, it will serve 2 million. A smaller plant operates in southern California. There are, geologists say, about 3 dozen other geothermal sites in the United States that could produce

electricity—and many more in other countries.

Just as the world will need new sources of fuel, it will need new sources of food. In 1975, the world population topped 4 billion. By the year 2023, the figure will have doubled. That means food production too must double. Anything less would spell disaster.

But only so much farmland exists. And that acreage shrinks every year, giving way to such things as erosion and the spread of cities. How then can the world increase its food supply?

"People are going to have to grow food by new methods and in regions such as deserts," says Carl N. Hodges, an expert in food production. "It's as simple—and as difficult—as that." Dr. Hodges directs the Environmental Research Laboratory (ERL), of the University of Arizona, in Tucson. ERL is a leading research center in the technology *(Continued on page 44)*

Geothermal food factory

Geothermal energy helps produce more than electricity. At Navarro College, in Corsicana, Texas, a great quantity of geothermal water lies underground. It is hot, but not hot enough to power machinery. It is briny, or salty, and cannot be drunk or used for watering crops. Still, the water helps produce generous harvests of shellfish and fish, fruit and vegetables.

Pumps bring the brine to the surface in pipes from a production well in rock layers deep below. In a heat exchanger, the brine heats—but never mixes with—city water. The brine, rid of its heat, flows into a disposal well, where it is reheated. The warmed city water flows into enclosed shrimp tanks. The shrimps grow rapidly in the heated water.

The water next flows in pipes into a greenhouse. It heats the building for growing tomatoes and other crops year-round. The crops are hydroponic—grown in enriched water alone. The heated city water is now lukewarm. It flows into an outdoor pond. In the pond, tiny plants called algae (AL-gee) thrive on nutrients, or food, contained in shrimp waste. Catfish cultivated in the pond feed on the algae and grow quickly for harvesting. The water returns through pipes to the heat exchanger, and the process begins again.

CITY WATER

SHRIMP

huehnergarth

FRUITS AND VEGETABLES

GREEN HOUSE

SHRIMP PONDS

POND WATER RECIRCULATED

Catfish Pond

CATFISH

HEAT EXCHANGER

GEOTHERMAL BRINE

PRODUCTION WELL 3,900ft

DISPOSAL WELL

Small fry mean big business on a new kind of ranch—a salmon ranch. At hatcheries, salmon eggs yield millions of fry. At six months, the fry are taken to ocean bay ponds, such as this one (above) at Coos Bay, Oregon. During a two-week stay, the fry register the special smell of their pond. Ranchers then release the fry to the ocean. Two or three years later, the fish, now adults (right), return to their home pond. The water's smell has guided them. Ranchers harvest the salmon. The fishes' eggs are gathered for the hatchery—and another round of easy fishing.

(Continued from page 41) of food production.

The lab lies in the hot, dry desert. If you visited its acres of greenhouses, you'd see some unusual sights. Spinach sprouts from revolving drums. Fast-growing fish called tilapia swim by the thousand in indoor tanks. Heads of lettuce, plugged into Styrofoam sheets, float on huge water troughs. It's all part of a new technology in food production. Scientists call the technology CEA, for controlled-environment agriculture.

CEA includes the raising of both plants and animals.

At ERL, technicians carefully control heat, humidity, light, and nutrition—everything that affects growth. "Water and nutrients [substances for growth] are both used over and over," says Hodges. "We throw away nothing and recycle everything."

Scientists at the lab try to find solutions to a wide range of food problems. They've developed a way for astronauts to grow their own vegetables in space. It's

44

FRANÇOIS GOHIER (BOTH)

Flower power: For purifying water, scientists are looking to the water hyacinth. They drop it into shallow sewage ponds (above). A few days later, they have pure water. Two containers (above, right) show the difference the plant makes. It absorbs dissolved wastes as nutrients. Bacteria at its roots eat solid wastes that fall to the pond bottom. In water-scarce San Diego, California, a million people will soon enjoy water purified by water hyacinths. There's more: Mature plants provide feed for livestock. Decaying plants produce fertilizer and give off a gas that can be used as fuel.

those spinach drums. The revolving motion would simulate gravity, which is essential to growth. A nutrient-rich mist, sprayed on the roots, would feed the plants. "It would cost roughly 500 dollars a pound to send fresh vegetables to the astronauts by shuttle," says Dr. Hodges. "The drums can cut the astronauts' vegetable bill down to a reasonable amount."

Desert areas, notes Dr. Hodges, cover a third of the earth's surface. "With CEA," he says, "people can grow crops in deserts and even in city spaces. They can grow great quantities of plants and animals to add to farm-produced food. At ERL we're producing up to a hundred times more food per acre than could be harvested from normal farmland or from the ocean."

Why don't farmers and fishermen around the world simply change over to CEA? For one thing, running a controlled environment takes special training. For another, CEA costs more than ordinary ways of raising food. That's especially true if the operation is a small one. Also, some foods—wheat, for example—are best produced in wide, open spaces.

Still, scientists say these new methods have much promise. "They won't solve the world's food problems immediately," says Noel D. Vietmeyer, another specialist in food production. "But they are a beginning. Undoubtedly, the technology can be made simpler and costs brought down."

Dr. Vietmeyer works with the National Academy of Sciences, in Washington, D. C. There are, he says, thousands of untapped sources of food. "Perhaps 20,000 kinds of edible plants exist," he points out, "yet only a hundred or so are grown as crops."

45

Held firm inside troughs, hydroponic lettuce grows clean and fast. Runoff water, rich in nutrients, is returned to the troughs to be taken up by the plant roots. Hydroponic farming helps fill the year-round demand for fresh fruit and vegetables.

Leaves overlapping, lettuce ready for harvest fills an acre of greenhouse troughs (left). These closely spaced lettuce plants grow hydroponically. "In New England, an acre of hydroponics in a year produces more than 50 times the food from a field acre," says lettuce grower Graham Davidson, of Salisbury, Connecticut. Here, he inspects his crop. Special equipment and the cost of heating fuel make greenhouses like this one expensive to operate. "Still," says Davidson, "hydroponic food prices keep coming down."

Foods of the future could include plants few people today know about. "Take amaranth," says Dr. Vietmeyer. "For centuries, Indians in Mexico and Central America ate this grain as a main part of their diet. They also used it, mixed with human blood, as a part of their religious ceremonies. This use shocked the Spanish conquerors, who came to the area in the 1500s. The Spanish made the Indians stop growing the plant. Amaranth was all but forgotten. Now we're rediscovering it."

Amaranth can grow in many climates and soils. You can pop it like corn, grind it into flour, make it into all sorts of food. It is richer in nutrients than just about any other grain. "Also—and this is very important—people like its flavor," says Dr. Vietmeyer. "It could be the grain of the future."

Another plant shows promise for widespread use as both food and fuel. It's a tree called leucaena (lew-SEEN-uh). The tree grows fast and tall. In only eight years, it can reach the height of a five-story building.

People in Asia and Latin America eat the leaves and seeds of the leucaena. The tree also provides fertilizer and a high-protein, low-cost cattle feed. People burn the wood for fuel. The wood can also be ground into pulp to make paper. The tree's bark and pods are used in making dyes. "Leucaena is now being seriously cultivated in several underdeveloped countries," says Dr. Vietmeyer. "It could soon provide those countries with food and fuel—and with badly needed money from exports."

There are still other ways to bring more food to people's tables. In the past few years, biologists have discovered how to change a plant's genetic structure—

47

Way to grow! At Walt Disney World Epcot Center, near Orlando, Florida, visitors can see new methods of growing food. In the background, hollow columns sprouting with vegetable plants move in a slow circle for even exposure to light. At left, lettuce grows on vertical surfaces. Saltwater plants, right, thrive in horizontal cylinders. The roots of all these plants touch no soil. They are sprayed or irrigated with nutrient-rich water.

the inner code that tells cells how to grow. Genetic changes can make a plant produce more and better quality food. A planned rearrangement of genes can cause a plant to grow taller, to resist disease, or to thrive in salty soil or in a very dry climate.

Using new knowledge, scientists can now multiply the number of cattle and other food animals. Suppose a certain cow produces unusually large quantities of milk. Dairy farmers the world over would want calves from her, all just like her. That's possible today. Only a few years ago it was not.

With chemicals called hormones, the scientists cause the cow to produce many embryos when she becomes pregnant. Normally, a cow produces only one embryo—the bundle of cells that will eventually develop into a full-grown animal.

A veterinarian flushes the dot-size embryos painlessly from the cow. Refrigerated, some embryos will live for two weeks. Frozen, they'll live two years, and perhaps much longer. The embryos are inserted in other cows. Those that survive develop inside their substitute mothers. In this way, one cow with prized

Tasty, too! In hydroponic greenhouses, ideal care lets plants do their best. Below, at Epcot, a type of large squash grows to maximum size.

A greenhouse at Navarro College (above) yields a huge crop of hydroponic tomatoes. In this controlled environment, vegetables grow quickly to uniform size and firmness (right bottom). Neither insect spray nor rough weather touches the crops. Geothermal energy keeps the greenhouse warm.

49

New breed of science: At Sungene Technologies laboratories, in Palo Alto, California, scientists create plants with specific desired traits. Above, embryos are removed from corn seeds. The embryos—plants in the earliest stage of growth— are placed in liquid food. By changing the food, scientists control cell behavior. Moved to soil, the embryos will mature with traits different from those of their parents. At left, a scientist examines an embryo for roots or shoots. At right, a researcher monitors sunflowers grown from controlled embryos. Scientists can also produce desired traits by removing and replacing genes— cell particles that hold instructions for growth.

traits can produce 16 calves, or even more, a year.

"In the future, the number will go much higher," says Alvin C. Warnick. Dr. Warnick works with the Animal Science Department of the University of Florida, in Gainesville. "Through a process called cloning," he explains, "we'll be able to produce endless numbers of fine cows—or most any other animal—all alike." (You'll discover in chapter 4 how cloning works.)

The technology for feeding a growing world population is now available. Some of it is costly. Some is highly complicated. Some will not be used for years to come. But for right now, there may be one simple and painless way to increase the supply of at least one essential food. Dr. Vietmeyer tells about it:

"Many Americans," he says, "simply won't eat a fish with the name 'roach' or 'ratfish' or 'shovelnose' or 'hogsucker' or a dozen others. Yet the flesh of these fish is perfectly delicious. In places where the fish have more appealing names, people enjoy them every day. Give the fish prettier names and—presto!—the United States could triple its annual harvest of fish protein."

It's worth thinking about.

Tomorrow's
TRANSPORTATION

At Walt Disney World Epcot Center, near Orlando, Florida, a train called a monorail whizzes past the exhibit Spaceship Earth. The single-rail train serves as a symbol of the future in transportation. Engineers are using new materials, new ideas, and new technology to build new types of vehicles. As a result, you may one day find yourself flying in a sun-powered airplane or skimming along in a train that floats above the rails.

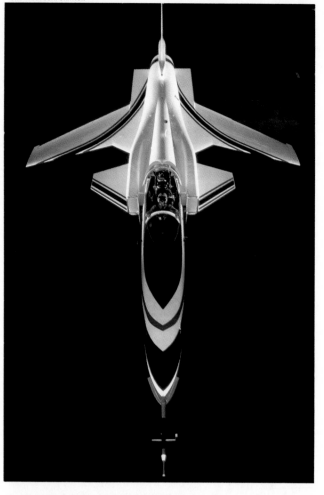

Teamed with humans, computers will copilot planes of tomorrow. One of those planes is the X-29 (left), an experimental U. S. fighter. In it, a computer (backed by two spares) makes 40 flight corrections a second. The forward-angled wings enable the plane to turn very quickly. The computer makes rapid adjustments that help keep the plane stable, or steady.

"A pleasure to fly." That's what pilots say about Beech Aircraft Corporation's upcoming Starship I (right), a business plane. Its rear-mounted engines reduce propeller noise in passenger compartment and cockpit. Tipsails mounted on the wings work like boat sails to give additional push. The small wings near the nose are called canards (kuh-NARDZ). They help make the plane safer to fly at low speeds.

omorrow, getting there could be a lot more than half the fun. To visit a friend across town, you might soar above the rooftops in your own lightweight aircraft. On a vacation trip, you could find yourself skimming along at 250 miles an hour (402 km/h) aboard a train that doesn't touch the rails. These are just two of the many exciting things that are happening in the world of transportation. The changes are helping people go places faster, with less expense, and often in greater comfort and safety.

Consider the modern airplane. In 1976, the first supersonic jetliner, the Concorde, began scheduled passenger service. The Concorde streaks over the Atlantic at 1,350 miles an hour (2,173 km/h), twice

the speed of sound. It flies from New York City to Paris, France, in only four hours. That's about half the time taken by standard jets such as the 747.

Designers want to do more than just make airplanes fly fast. For example, they're using new plastics and metals—strong but lightweight—to make planes more fuel-efficient. They're changing the shapes to make planes more maneuverable—able to turn and climb quickly in a small area.

One new experimental plane, a fighter, looks as if it were flying in the wrong direction. The craft, called the X-29, has wings that extend slightly forward. The angle makes the plane highly maneuverable at supersonic speeds—with the aid of computers. Without the computers, flying the X-29 would be like racing a horse on a sheet of ice.

"The plane requires (Continued on page 59)

Look, Ma, no fumes: The Solar Challenger *(above) flies on sun power alone. Solar cells on wings and tail convert sunlight into the electricity that turns the propeller. Paul MacCready, of Pasadena, California, built the plane in 1980 to prove that solar-powered flight was possible.*

Long time aloft: The Voyager *(left) is set to fly around the world nonstop—without refueling.* Voyager's *designer, Burt Rutan, of Burbank, California, made the plane extra lightweight and fuel-efficient.* Voyager *will carry a pilot and copilot. It will cruise at 100 miles an hour (161 km/h). During the 12-day flight,* Voyager *will burn only 1,500 gallons (5,678 L) of fuel. The flight is scheduled for late 1985.*

Build your own airplane? It's a possibility. Kits for do-it-yourselfers bring the price of a personal plane down to about the cost of a medium-priced car. Here, George Edwards, of Cupertino, California, assembles a small plane in his garage.

57

Red Baron, watch out! His hand on the control stick, Ben Sclair, 13, guides his ultralight through the sky. Ultralights—simple, one-person airplanes—are turning people of all ages into pilots. The craft can take off and land in a medium-size backyard. "Flying an ultralight is peaceful—but exciting," says Ben, who lives near Tacoma, Washington. Ultralight pilots do not need licenses, but Ben took flying lessons anyway—as should anyone who takes to the air.

(Continued from page 54) about 40 flight corrections every second," says Robert P. Harwood, an official with the Grumman Corporation. Grumman, with headquarters in Bethpage, New York, builds the X-29. "No human pilot is quick enough to make all those corrections," Harwood adds.

The pilot works the steering controls as he would on a regular fighter plane. A computer translates the pilot's commands into exact adjustments. "Without those constant adjustments," says Harwood, "the pilot could not keep the plane under control. It would shake apart in a split second."

Computers also help run the world's largest machines: giant oil tankers. The largest tanker, the Liberian-registered *Seawise Giant,* is five times longer than a football field. You might think the ship would need an extra-large crew to handle it. It doesn't. Thanks to computers that do much of the work, the crew numbers only 36. Much smaller, older ships might have crews of as many as 60.

Using satellite transmissions, computers continuously pinpoint the ship's location. They keep track of the systems that operate the ship's electricity, plumbing, engines, and automatic steering. They even calculate the best order of filling and emptying the ship's many huge oil tanks. Tanks filled or unloaded in the wrong order might make the ship lean to one side.

Just as planes and ships are changing, so is the

WHEELBAS
OAH
OAL
OAW
TR

automobile. The first automobiles were simply small carriages with motors. Over the years, cars became larger and heavier. Their engines became more powerful. Gasoline was cheap. Three dollars might pay for a fill-up. Most people could easily afford the large amounts of fuel required for big, powerful cars.

Then, in the 1970s, the price of gasoline rose sharply. That caused designers to seek ways to make cars cheaper to operate. One solution: Make them smaller. General Motors, the largest U. S. automaker, has built an experimental two-passenger commuter, the TPC. The TPC gets up to 95 miles a gallon on the highway. That's about twice the mileage of any other car made today.

Designers have also streamlined cars for better performance. At highway speeds, some cars burn more than half their fuel in fighting drag, or air resis-

tance. Many of today's cars have sleek shapes that slice through the wind like sharp knives. They cut gasoline use by as much as 20 percent.

Computers, of course, play a large part in making cars more efficient and, in some cases, easier to repair. Computerized fuel systems, for example, can help increase gas mileage by up to 50 percent. Volkswagen is using computers to help spot mechanical and electrical problems. A mechanic just wires a car to a computer that analyzes each system. The computer prints out a complete report of the car's condition. The report notes if anything needs repair.

Carmakers are developing dashboard navigation systems. A computer keeps track of the car's exact location. On a video map, it shows the driver the best route through a city or between one city and another. A Japanese company is experimenting with a car

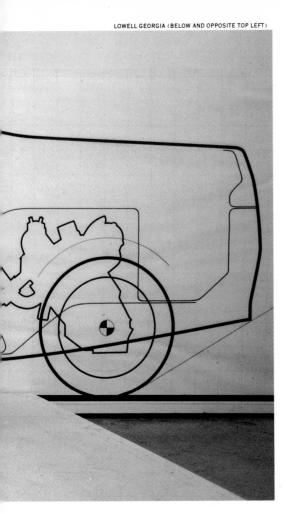

At the Art Center College of Design, in Pasadena, California, students try out a seating arrangement they designed (above). Instructor Richard Hutting notes their reactions. Low, sleek styling, such as that outlined here, reduces drag, or air resistance. The result: better fuel efficiency.

To get about town in the future, you may use an HPV—a human-powered vehicle. Four people pedal the HPV Pegasus (right). It cruises at 25 miles an hour (40 km/h). Students at the University of Cincinnati, in Ohio, designed Pegasus for speed and safety.

Cars of the future? Art Center students discuss sketches for a 14-week auto-design project (below). Students try to solve design problems—for example, seating that is hard to get in and out of. Proposed solutions to that one: swivel seats, and doors that swing up instead of out.

Researcher David Warn works at the Autocolor system he developed. Autocolor allows designers to create three-dimensional images on a computer screen. A quick and easy system, it may eventually replace hard-to-make clay models. Warn works at General Motors Research Laboratories, in Warren, Michigan.

61

TRANSRAPID INTERNATIONAL, GESELLSCHAFT FÜR MAGNETBAHNSYSTEME

Find a substitute for the wheel? Unthinkable—except that it has been done. Trains that ride on a cushion of air instead of on wheels now transport people in areas of West Germany, Britain, and Japan. Powerful electromagnets lift the trains and pull them along at 250 miles an hour (402 km/h)—or even faster. The German model (left) operates by magnetic attraction. A Japanese version (right) uses magnetic repulsion—a pushing away. Magnetic trains, called maglevs, may be introduced in the United States in the 1990s.

that's equipped with radar. If the car gets too close to a large object, it automatically stops.

Years before automobiles came along, trains were carrying people and freight in many parts of the world. Trains still are a major form of transportation. They haul enormous loads. They are more fuel-efficient than cars, trucks, or airplanes.

The world's two fastest standard trains speed along at about 150 miles an hour (241 km/h). They ride on rails in Japan and in France. Now engineers have developed a new type of train. It can go nearly twice as fast as the fastest trains today. Riders hear only a gentle *shhhhhh*. This train does not ride on rails. Instead, it rides on a cushion of air.

The new trains are called maglevs. They work through magnetic levitation, or lifting. Powerful electromagnets raise a maglev train half an inch (1 cm) above its guideway, or track, and pull it forward. Since no metal-to-metal friction results when it moves, a maglev train can reach much greater speeds than a standard train.

"Riding on a maglev is like flying at ground level," says Paul Sichert, an official with The Budd Company. Budd, located in Troy, Michigan, plans to manufacture maglevs for travel in the United States.

Right now, maglev trains are being built in the Federal Republic of Germany and in Japan. "We expect to see maglevs in operation in the United States by the early 1990s," says Sichert. Maglevs may make it easy to live in one city and work in another hundreds of miles away. A maglev could whisk a worker from Washington, D. C., to New York City, for example, in less than an hour.

"Maglev trains have advantages other than speed alone," says Sichert. "They use energy efficiently. They don't wear out. They have built-in collision safeguards. And they can operate in any kind of weather."

Trains, cars, ships, and planes—they're all changing as we find new ways to travel more rapidly, safely, and conveniently. In the world of transportation, nothing stands still.

Safe landing: Pilot and copilot prepare to touch down at a busy big-city airport—or so it seems. Actually, they haven't left the ground. They're training in a simulator. It copies the sights, sounds, and feel of a Boeing 757 cockpit down to the last detail. Using simulators, pilots can practice landing in all kinds of weather and in any traffic condition without risking aircraft or lives.

A captain steers a ship up a tricky channel (below). Red instrument lights aid his night vision. The night, black and moonless, holds danger of running aground or hitting another ship. But if such a thing should happen here, no damage would result. The captain is training in a simulator. It's at the Maritime Institute of Technology, in Linthicum Heights, Maryland.

N.G.S. PHOTOGRAPHER JOSEPH H. BAILEY (ALL)

In a driver-training class in Oradell, New Jersey, students test their skills in simulators. With simulators, students can experience what it will be like to drive through storms, over ice, in bumper-to-bumper traffic, and across narrow bridges. Simulators provide safe, effective practice for the real thing.

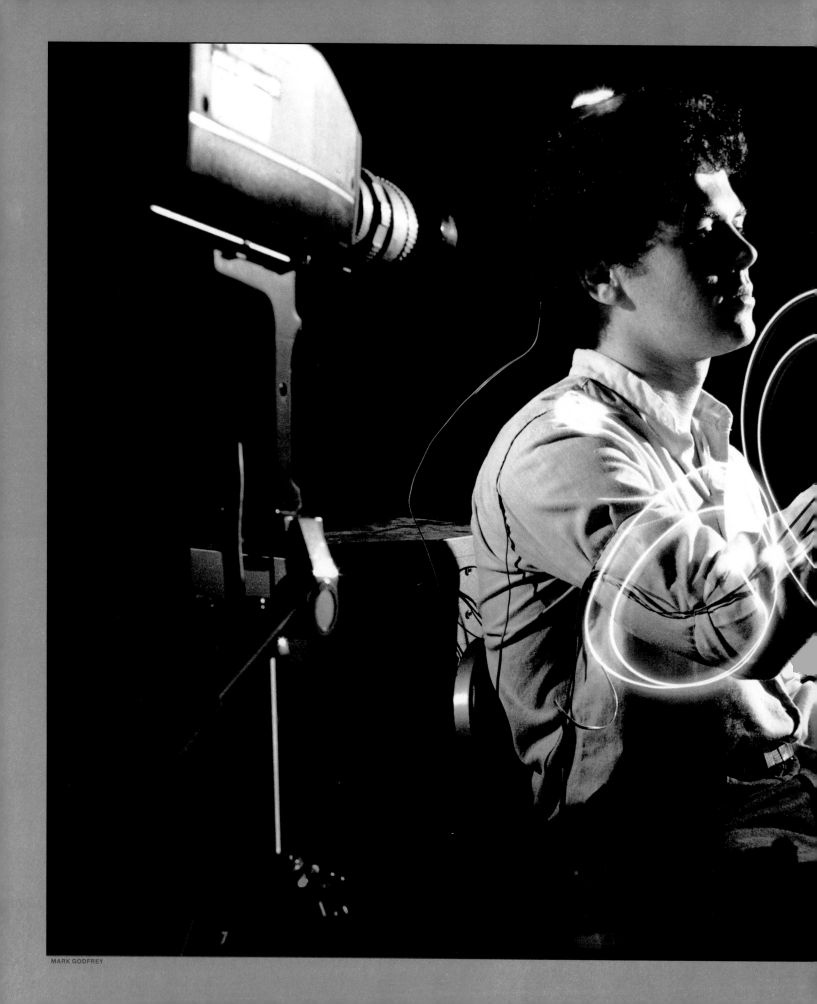

MARK GODFREY

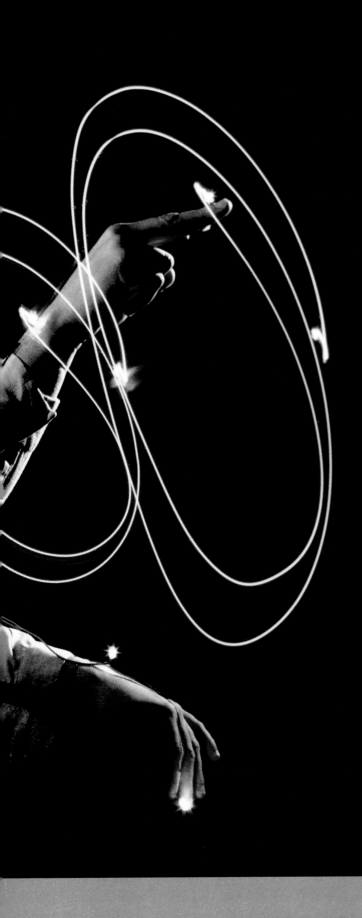

Breakthroughs in
MEDICINE

Doctors now fight disease and disability with space-age tools and knowledge. Here, James Tucker, of Fremont, California, forms sign-language phrases with lights. James has been deaf since birth. A video camera tapes his motions. The patterns will be stored as numbers in a computer. Scientists will analyze the patterns, reconstructed as lines on a video screen. The information will help the scientists learn how the brain processes language symbols.

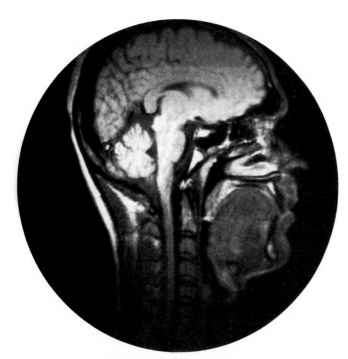

A look inside the brain—that's what a machine called a magnetic resonance (MR) imager provides (left). Cell centers, lined up by magnetism, swing back and forth to on-off radio signals. The cells absorb, then release, the signals. A computer records the signals and makes a picture from them. A series of MR scans can be stacked to show the whole organ.

Guided by a laser beam, a hospital technician positions a patient for an MR brain scan (right). The imager "sees" through the skull. Doctors use MR scans to locate damaged or diseased tissue. MR and other scanning techniques provide more information than ordinary X-ray pictures do.

friend of yours, call him Sammy, suddenly becomes ill. Sammy's mother takes him to the hospital. A doctor examines him and runs a few simple tests. Then the doctor turns to a computer and types in his findings.

Instantly, several questions appear on the screen. The doctor checks Sammy some more and enters his new findings into the computer. The computer tells the doctor what may be wrong with Sammy. The doctor is relieved. He and the computer agree. They diagnose, or identify, the illness as a touch of the flu—uncomfortable, but not serious. The doctor sends Sammy home with instructions for bed rest.

Within the next few years, Sammy's story will become increasingly routine. Though they don't wear white coats, computers are doing important work in medicine. More and more physicians rely on computers to help diagnose their patients' illnesses and to help determine the best treatments.

Today's doctors have to study constantly to keep up with rapid changes in medicine. Experts estimate that a physician must have at hand up to two million medical facts. To memorize all those facts would be impossible or close to it. But a computer can store them, and a doctor can add facts as they become known. By tapping a few keys on a computer terminal, the doctor can call up and sift through any information he or she needs. It takes only moments.

That's just the beginning of the computer's role in medicine. Linked with machines called scanners, computers help "image," or make pictures of, the inner parts of the body.

"For centuries," says Dr. Martin J. Kaplitt, a heart specialist, "physicians have dreamed of peering inside the human body without cutting into it. Ordinary

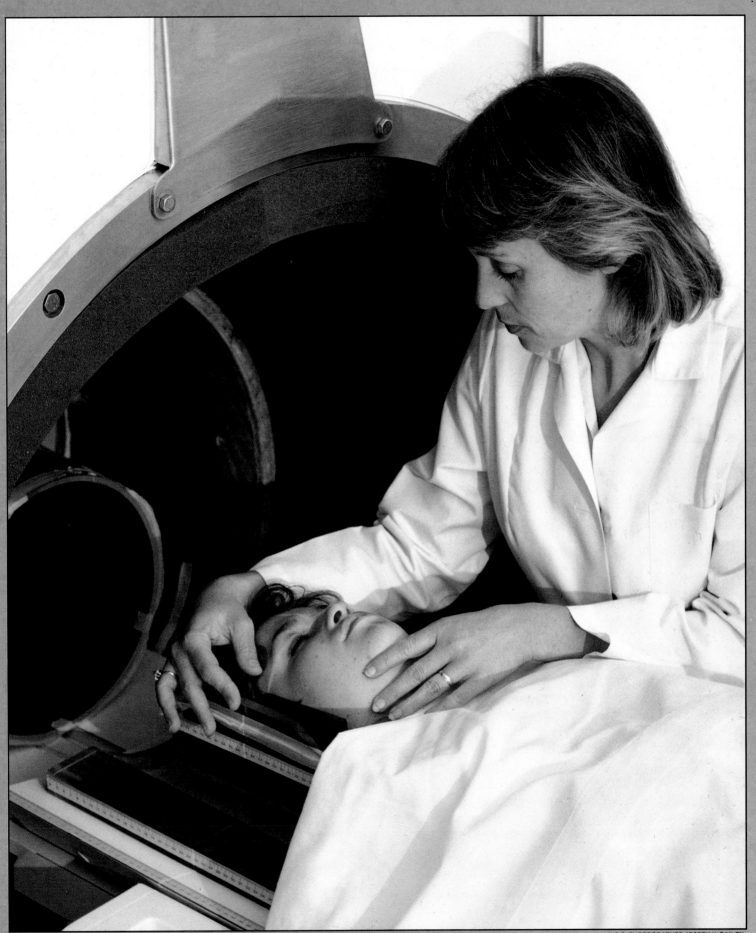

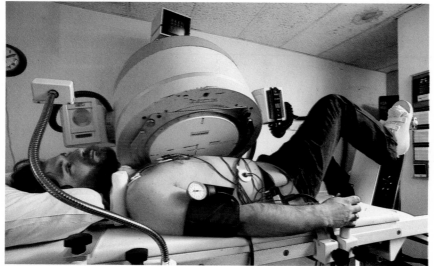

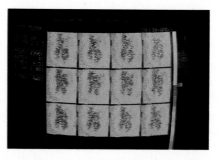

Using sound waves, a clinic technician scans the inside walls of an artery leading to the brain (right). The technician holds an ultrasound transmitter. Its high-frequency waves bounce off interior walls of arteries. The echoes produce an image. In a matter of minutes, ultrasound can scan the two main arteries to the brain. Computer images show whether an artery is clogged.

As he works pedals, this "bicyclist" has a gamma scan made of his heart (above). Minutes earlier, a nurse injected a radioactive material into the man's bloodstream. The substance has collected in the heart and now gives off low-level radiation. The radiation is detected by the scanner above the heart. A computer changes the radiation pattern into an image. At right, a series of images shows a heart at different stages of its pumping action.

X rays were a start, but they have disadvantages. For example, they are good for studying broken bones, but can be used only with difficulty in studying other organs. Modern technology solves that. We can say that our dream is finally coming true."

Dr. Kaplitt practices at Vascular Diagnostics, a clinic in Flushing, New York. During an examination, he rubs a device against the neck, abdomen, and legs of a patient. The device, an ultrasound transmitter, sends high-pitched sound waves into the patient's arteries. A computer helps change the returning echoes into images. They appear on a computer screen like a black-and-white movie.

"With ultrasound," says Dr. Kaplitt, "we can examine the inside walls of the two main arteries to the brain. It takes just a few minutes. The pictures let us see if deposits of fat are starting to block the passages. If they are, we can treat the patient and prevent major illness such as stroke."

A doctor can do an ultrasound scan in the office. More complicated scans usually take place in the hospital or in a scanning facility. To view a cross section, or slice, of the body, a doctor may use a computerized axial tomography scan. It's called a CAT scan, for short. Most medical centers today have these machines. A CAT scanner uses X rays to help create an image. Unlike an ordinary X-ray picture, however, a CAT scan appears on a video screen. A computer colors the image. A CAT scan shows muscle, bone, and all other types of tissue. By contrast, an ordinary

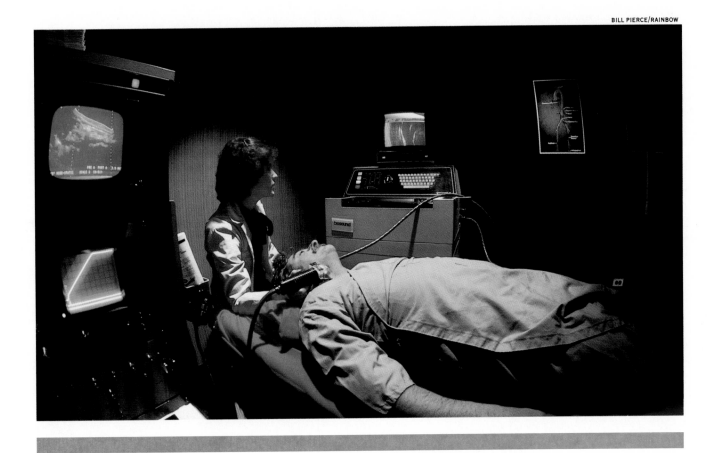

X-ray picture shows only the shadows of bones.

To watch an internal organ at work, as well as to see what the organ actually looks like, doctors use a positron emission tomography scanner. Just call it a PET scanner. A doctor first injects a patient with a solution that contains a small amount of radioactive material. The formula varies with the organ to be scanned.

When the solution reaches the organ, a reaction takes place. Particles in the solution crash into particles in the body cells. The collisions produce tiny rays. Detectors outside the body pick them up. A computer converts the rays into a PET scan showing slices of the organ.

"The computer can stack up the slices to show an entire organ in 3-D," says Dr. K. Lance Gould, of the University of Texas Health Science Center, in Houston. "I can have the computer 'slice' a heart in half to let me see how the blood flows through the interior. I can watch the heart beat. Computer coloring shows me which areas are abnormal. The scan allows me to see the heart from different angles. It's as though I could take the heart out of the patient's body and hold it in my hands."

Another type of scanner looks inside the body with magnetism and radio waves. The machine is a magnetic resonance, or MR, imager. Its strong magnetism holds the nuclei, or cores, of atoms in a uniform position. The scanner directs radio waves at hydrogen atoms, which are present in all body cells. The nuclei absorb, then give off, the waves. A computer makes

71

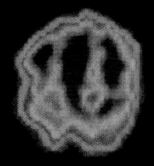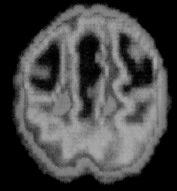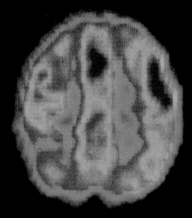
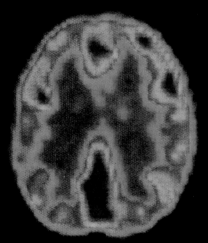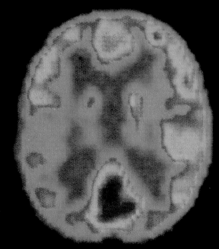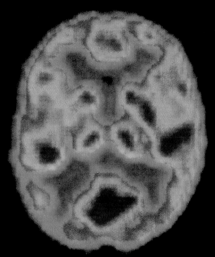
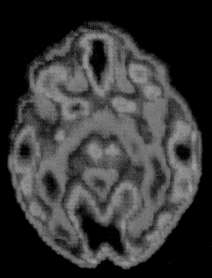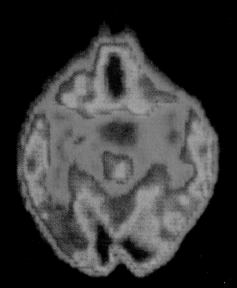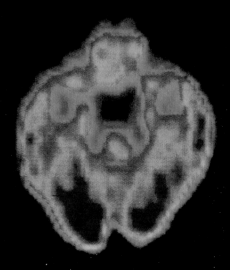

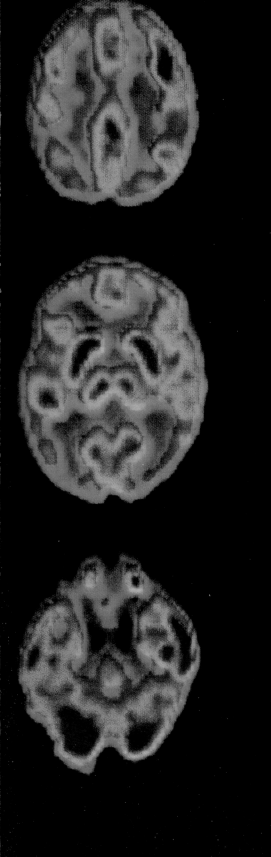

Layer by layer, PET scans reveal the brain at work. These scans read from left to right. PET scans show which areas burn up glucose— the brain's fuel—faster than others. Red shows areas of highest activity; blue, areas of lowest activity. The person being scanned was listening to a reading of a Sherlock Holmes mystery. A red splotch at either side of scan seven shows that the person's hearing centers were active.

an image of those given off. The nuclei in diseased cells do not respond as those in healthy cells do. A doctor can easily spot them in the computer image.

Modern machines that "see" inside the body all use computers to make sense of the signals they receive. Computers are also becoming essential to bionic medicine—the replacement of body parts with scientifically engineered parts.

You may have seen reruns of TV's "The Bionic Woman." Severely injured in a skydiving accident, the heroine was rebuilt with artificial parts that gave her superhuman strength. That's still science fiction. But specialists can build bionic parts that look —and work—much like the real thing.

Arms, knees, elbows, wrists, skin, heartbeat regulators, blood vessels, and the heart itself—now these can often be replaced if lost to illness or accident. And the list keeps growing.

Artificial body parts have been around for centuries. They began with the peg leg and the metal hook. But only since the 1970s have scientists developed legs and arms that work electronically, with motions resembling those of real limbs.

One of the most advanced bionic limbs is the Utah arm. Scientists at the University of Utah, in Salt Lake City, developed it. Small metal disks called electrodes are taped to the skin above the Utah arm. The electrodes pick up (Continued on page 76)

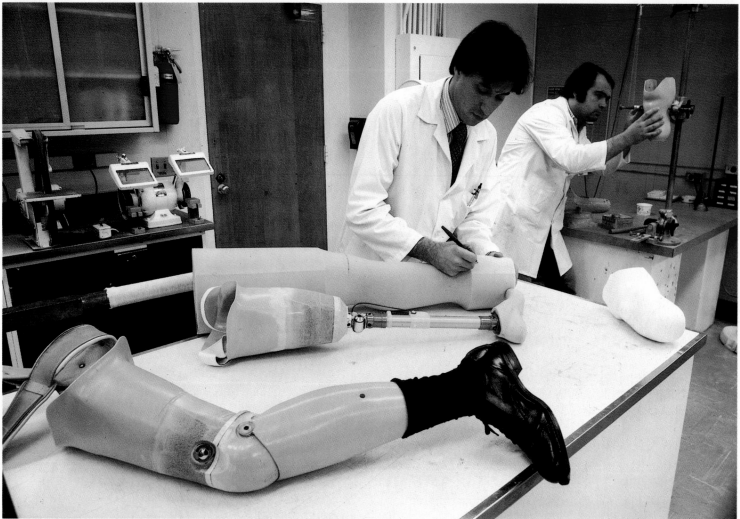

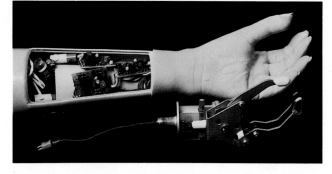

Making a leg that matches its mate calls for careful workmanship (above). In a Rusk Institute lab at New York University, in New York City, an expert in bionic parts marks a plastic ankle for exact trimming. Another technician works on part of a kneecap. Artificial limbs will never work as well as natural arms and legs. But today's bionic limbs come closer to it than anything yet.

Batteries, a minicomputer, and a motor make up the inner workings of this bionic arm. The arm appears lifelike, and it moves naturally—a big advance over the mechanical hooks used only a few years ago. The metal "skeleton" of the hand shows below the arm.

Finishing touch: Plastic skinlike gloves (right) fit over the outer casing of bionic hands and arms. The gloves are tinted to match the wearer's skin. Fingerprints and palm lines make the hands look even more real.

74

In a few minutes at a computer, a technician (left) finds the correct size and shape for an artificial wrist joint. This work is taking place at the Hospital of Special Surgery, in New York City. The computer turns an X-ray view of the wrist bones into a three-dimensional pattern. It shows exact measurements. The result: a joint that fits perfectly (right).

(Continued from page 73) tiny electrical signals from the muscles and send the signals through wires to the arm's electronic controls. The signals make the bionic arm move. A minicomputer and a tiny, battery-powered motor inside the arm help it move smoothly.

And what about help for the blind? In recent experiments, electrical signals were sent through electrodes to the vision center of a blind person's brain. The electricity stimulated, or excited, the vision center to "see" points of light. Then signals forming Braille symbols or letters of the alphabet were sent. The blind person could read them. "It's a start," says Dr. Richard Norman, of the University of Utah. That's where the experiments were conducted. "Someday," he adds, "it may be possible to transmit pictures, too."

Deaf people can have receiver-sender electrodes implanted in bone behind the ear. Wires transmit sound from the electrode to nerves in the inner ear. "The wearer can hear many attention-getting sounds, such as dogs barking, bells ringing, sirens wailing, and voices," says Dr. William H. McFarland. He's a

hearing specialist at Gallaudet College for the Deaf, in Washington, D. C. "The wearer cannot yet make out spoken words," Dr. McFarland adds. "That will be the next big step."

Vital organs too are entering the bionic age. Vital organs are organs essential to life. They include lungs, kidneys, and heart.

For about 20 years, doctors have been performing heart transplants. The patient receives the heart of a donor, a person who has recently died. At first, the transplants allowed the patients to live only a little longer. The patients' bodies then rejected the foreign organs. Now scientists have developed new drugs that help prevent rejection.

Before long, donors may not be necessary. Bionic devices may be able to replace vital organs. "The technology is here," says Dr. Arnold Landé, a heart surgeon and medical inventor in Houston, Texas. "We are making workable organs in the laboratory right now. The next step is to have them approved for general use."

Dr. Landé has designed a bionic heart-lung. He is

76

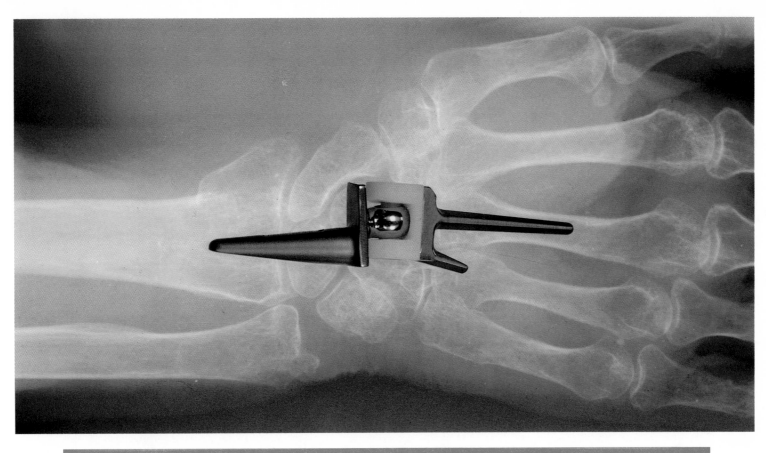

now testing a bionic kidney. "Any new medical device must be tested over and over," he says. "Then it must undergo review by a series of medical boards and government agencies. The process takes time and money. But there's good reason for such caution: Human lives will depend on the devices."

Amazing as they are, transplants and artificial parts can't solve some problems, such as paralysis. Some people lose the use of their legs because messages from the brain do not reach the legs. Such people may have had a stroke or injured the nerves that carry the messages.

A team of government doctors and engineers is tackling the problem of paralysis. In a project at the Veterans Administration Hospital in Cleveland, Ohio, the team has designed a special computer the size of a small camera. The computer, programmed with leg movements, is linked by electrode wires to nerves in the legs. It substitutes for the brain in giving commands to the legs.

"The patient pushes a button for the function he needs. The computer tells the nerves and muscles exactly what to do," says Dr. Ernest Marsolais, who directs the program. "Already, six patients are walking. They're also climbing stairs."

When will the breakthrough benefit others? "The leg computer will be generally available by the end of the 1980s," Dr. Marsolais predicts.

Successes in bionic medicine are being matched by those in surgery. For hundreds of years, surgeons have used delicate knives called scalpels when they have had to cut into the body. Now they have a new tool: the laser. Many doctors consider it far superior to the scalpel.

A surgeon can use a laser beam to vaporize diseased tissue. "The tissue just goes up in smoke," says Dr. Mark S. O'Brien. Dr. O'Brien practices at the Emory University School of Medicine, in Atlanta, Georgia. He specializes in treating tumors in children.

"The laser beam I use vaporizes unhealthy tissue," Dr. O'Brien continues. "It affects only a few millimeters of tissue at a time, so it's easy to control. Using a

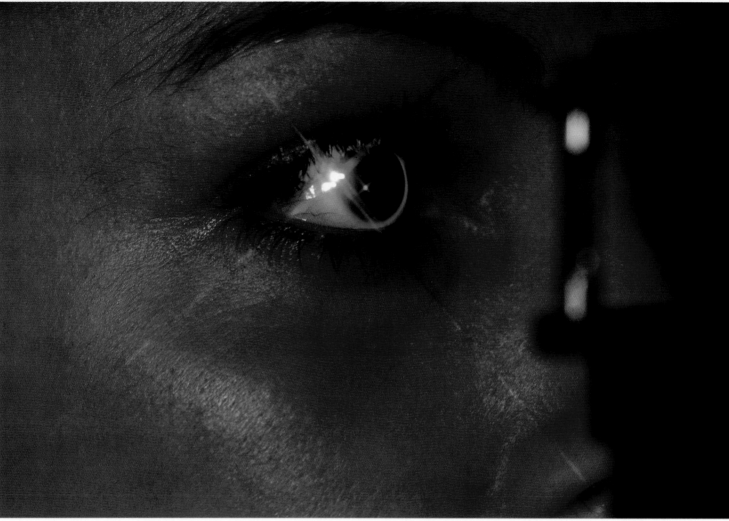

Right on target, laser light repairs a torn retina, tissue at the back of the eye. The operation takes only minutes. It does not hurt. Heat from a laser beam can also seal bleeding ulcers, unclog arteries, and weld torn nerves. Doctors predict that lasers will replace knives for most surgery within a few years.

laser with an operating microscope, I can remove all of a tumor. Surrounding tissue, untouched by the laser beam, is left unharmed."

Lasers have other medical uses. The beams reconnect cut nerves. They remove birthmarks, warts, and other blemishes. They mend delicate torn tissue, as in eye injuries. Sent through an optical fiber, they clear blocked arteries.

Other exciting advances are taking place in microbiology. Let's return to your friend Sammy. Suppose

the doctor came up with a different diagnosis. Instead of flu, the illness turned out to be strep throat. It's a painful infection that can become serious.

The standard treatment would be penicillin or some other antibiotic. But in a growing number of hospitals and clinics today, the doctor would prescribe antibodies. You may have read about antibodies in YOUR WONDERFUL BODY!, another volume in this series of Books for World Explorers.

"Antibiotics" and "antibodies" may sound similar,

78

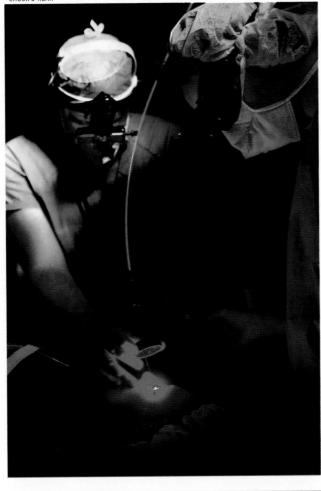

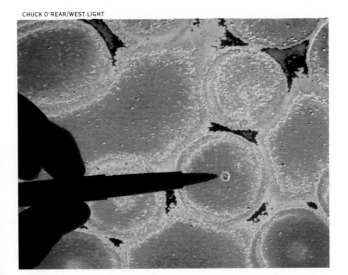

Super small: A laser beam can be made narrow enough to drill a hole in a red blood cell (above). The hole appears here greatly magnified. A human hair is about 160 times wider.

Light fantastic: Surgeons at Children's Hospital in Los Angeles, California, use a laser beam to vaporize an eye tumor. An optical fiber carries the beam. Narrow and accurate, the beam does not touch healthy tissue only a few cells away.

but there are important differences. Antibiotics are chemicals made by tiny organisms such as fungi and bacteria. Antibiotics poison many kinds of germs—both dangerous *and* helpful. Antibiotics have saved countless lives. But they can have unpleasant or dangerous side effects.

Antibodies are living cells produced by the body itself. There are thousands of kinds of antibodies. Each kind fights a specific, harmful kind of germ or chemical. Antibodies cause no side effects.

Now scientists have succeeded in growing great quantities of antibodies in the laboratory. They force two cells together into one new cell. One of the two starter cells is an antibody producer. The other is a type that reproduces itself rapidly.

The new cell does two jobs: 1) It produces antibodies and 2) It reproduces itself in endless numbers. Scientists have a word for this endless, exact reproduction of a single cell. The word is "cloning." Antibodies produced in *(Continued on page 82)*

79

A world of medical wonders

Almost every day, medical scientists come up with something new. On these pages, you see some of the latest inventions. **1.** Using strips of tape called Dermazip, doctors zip a wound closed instead of sewing it. **2.** MedTymer won't let you forget to take a pill. A battery-powered cap flashes and sounds an alarm at medicine time. **3.** New pills act as pumps after you swallow them. Body water seeps through a pill's plastic skin. Pressure builds inside the pill. Through laser-drilled holes, the pill releases a steady flow of medicine for 24 hours. **4.** This seasickness medicine isn't hard to swallow. It's contained in a small patch that sticks behind the ear. The patch slowly releases medicine that is absorbed through the skin. **5.** Space

JOHN HUEHNERGARTH

helmet? No, pollen-proofer. Allergy sufferers wear the Hincherton Hayfever Helmet to filter the air they breathe. **6.** Implanted under the skin, a long-lasting solid plastic capsule will deliver a steady amount of insulin to people who have diabetes. The device, now being developed, will replace daily injections. **7.** Computers can store and process millions of bits of information. They have dozens of uses, from helping doctors diagnose illness to helping other machines look inside the body. **8.** Sun Timer tells you when your skin has had enough sunshine. The light-sensitive badge changes color when it's time to cover up. It can help prevent sunburn and skin cancer.

Aided by a computer, John Duganne learns a new way to write (above). John, 13, has cerebral palsy. He cannot control his hands. He speaks with difficulty. Here, Dr. Laura Meyers works with him on the computer. She helped write a computer program that makes writing easier for John. He uses a head wand to type (above, right). A computer voice says each letter as he types it. When John taps a special key, the computer reads what he has written. A printer turns out a copy of his work. As he writes a story (right), John and Dr. Meyers share a laugh. John attends John Adams Junior High School in Santa Monica, California. He takes his computer to school on a motorized wheelchair. John's goal? "To become a professional writer," he says.

(Continued from page 79) this way are called monoclonal (MAHN-uh-KLONE-uhl) antibodies.

Some germs attack the body much faster than it can make antibodies. And some people do not make certain antibodies at all. Monoclonal antibodies can be injected in large quantities into their bloodstreams. There, the monoclonal antibodies overwhelm the invaders. They work safely as a natural defense.

Scientists have another way of engineering cells to work for people. They change the instructions a cell contains. The instructions are carried by the genes. Every cell holds thousands of genes. They are grouped in an arrangement that looks somewhat like a spiral staircase. Scientists now know how to cut a selected gene from the staircase. The scientists then fasten a different gene in the empty space. The technique is called gene splicing.

The new gene tells the cell to produce a certain chemical. By cloning the new cell complete with its spliced gene, scientists can produce a harvest of natural chemicals vital to the human body.

In this way, scientists can mass-produce human insulin, to take one example. Insulin is a chemical that helps the body use sugar and starches. Without insulin in the bloodstream, a person would die. Most people produce needed amounts of insulin naturally, but

82

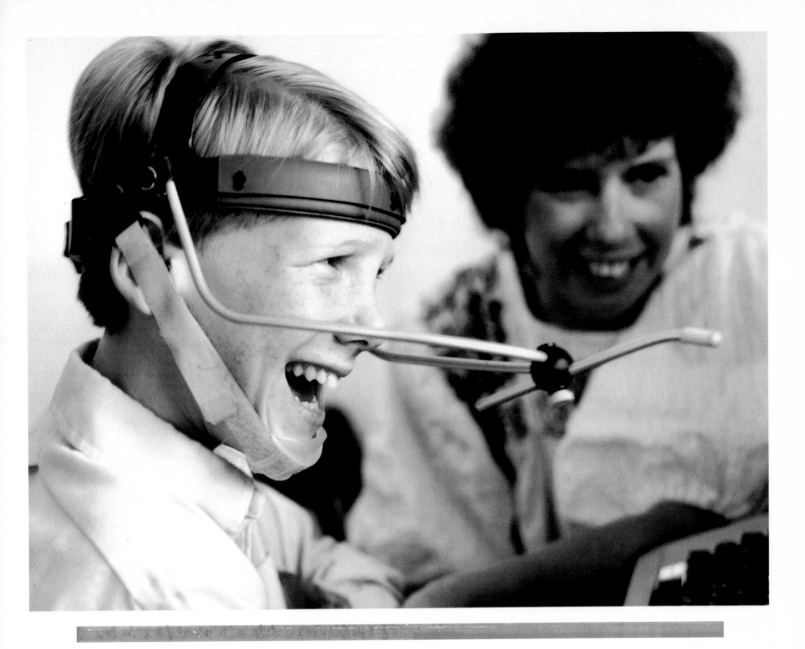

those who have diabetes do not. Insulin injected into a diabetic's body saves the person's life.

Genetic engineers are working on new treatments for other diseases as well. In early 1985, researchers announced that they had developed an antibody that fights a form of multiple sclerosis in laboratory mice. The antibody captures malfunctioning white blood cells that are eating away tissue that protects nerves. Other scientists have developed substances that instantly dissolve blood clots in humans. The substances could halt thousands of heart attacks every year. Scientists also have developed antibodies in the laboratory that recognize and destroy some types of cancer cells.

How much can biological engineering accomplish? No one knows for sure. The science is still young. Many scientists think it could, in time, wipe out most diseases. There's no question about one thing, however: Together with other breakthroughs in medicine, the new biology is helping people lead healthier, more comfortable, and longer lives.

83

5

SPACE

New Worlds To Discover

Once, the big challenge was just getting there. Now astronauts travel into space routinely. Here, Mission Specialist Bruce McCandless II goes on a space walk. To move about, he uses an MMU— a manned maneuvering unit. On each mission, astronauts make important discoveries. Together with scientists on the ground, they are opening up the universe in what is perhaps the biggest adventure ever undertaken.

Suited for space, Mission Specialist Robert L. Stewart moves outside the shuttle Challenger *(left). There is no air in space. Stewart's sealed suit provides oxygen, a comfortable temperature, and atmospheric pressure. Without the pressure, an astronaut's body would blow up like a balloon. Inside Stewart's helmet, a fruit bar is positioned for easy nibbling. There is also a water supply. To guide his movement, Stewart activates small jets on the MMU.*

Running tied down gets Mission Commander Joe H. Engle nowhere—and that's just the way he wants it (right). Astronauts must exercise since muscles weaken rapidly in weightless space. The elastic straps keep Engle from floating off. They also provide resistance that helps keep his muscles in tone.

At dawn on July 20, 1969, three men put on their suits and went to work. These men weren't headed for another day at the office, however. Neil Armstrong, Edwin Aldrin, and Michael Collins were astronauts aboard the Apollo 11 spacecraft. They were getting ready to make the first manned moon landing in history.

When Armstrong set foot on the moon that day, after a 3½-day flight through space, he described the achievement as "one small step for a man, one giant leap for mankind." The first walk on the moon represented a major advance in the exploration and practical use of space.

Other advances followed. Today, hundreds of artificial satellites orbit the earth. They do a wide assortment of jobs. Communications satellites relay TV and telephone signals around the world. Weather satellites give early warnings of storms. Since the first one went into orbit in 1960, weather satellites have saved tens of thousands of lives on several continents. Military satellites keep watch on troop, ship, and aircraft movements. Some satellites are equipped to pinpoint downed planes and ships in distress.

Space exploration touches your daily life. Do you have a pocket calculator? A digital watch? These handy items resulted from research done in the space program. Solar energy cells also came about largely as a result of space research. Miniature power systems developed for use in space are saving lives on earth. The tiny packs run electronic pacemakers, devices that keep faulty hearts beating regularly.

"People are benefiting in countless ways from the space program," says William J. O'Donnell, a spokesman at National Aeronautics and Space Administration (NASA) headquarters, in Washington,

Preparing to land, Mission Specialist Ronald McNair (above, at left) checks his oxygen equipment before putting on his helmet. Stewart's helmet is in place. Connected by hoses to tanks, the helmets fill with pure oxygen. Like football players after hard play, astronauts breathe oxygen to stay alert. They wear helmets only during takeoff, landing, and space walks.

Housekeeping, space style: Jack Lousma, a mission commander, retrieves containers of lithium hydroxide from storage bins. The chemical solid acts as a filter for the shuttle's air-circulation system. It removes odors and carbon dioxide, a waste product in exhaled breath.

D. C. "In fact, a Congressional study shows that the benefits of space exploration have outweighed the costs by 14 to 1."

In 1981, a space shuttle, *Columbia,* soared into orbit for the first time and returned safely to earth. Earlier launch vehicles could be used for one shot only. Shuttles can be used again and again. A shuttle can put a satellite into orbit more cheaply than it can be launched from earth. Scientists on shuttle missions do experiments that cannot be done on the ground.

Businesses and governments are lining up to reserve cargo space and astronaut time for shuttle flights. They want to have special-purpose satellites launched. They seek scientific knowledge that can

be gained only in space. *Columbia,* along with three sister ships—*Challenger, Discovery,* and *Atlantis*—will make dozens of flights in the coming years. Altogether, the shuttles will carry hundreds of men and women. You could be a crew member. Living aboard a shuttle craft will be a lot different from anything you've ever experienced.

In space, you'll have to get used to weightlessness, the absence of gravity. Astronauts call the condition "zero g."

"I spent a lot of time enjoying zero g," says Robert L. Crippen, who piloted *Columbia* on the first voyage. "At first, I did things that surprised me, like shoving off from one side of the *(Continued on page 92)*

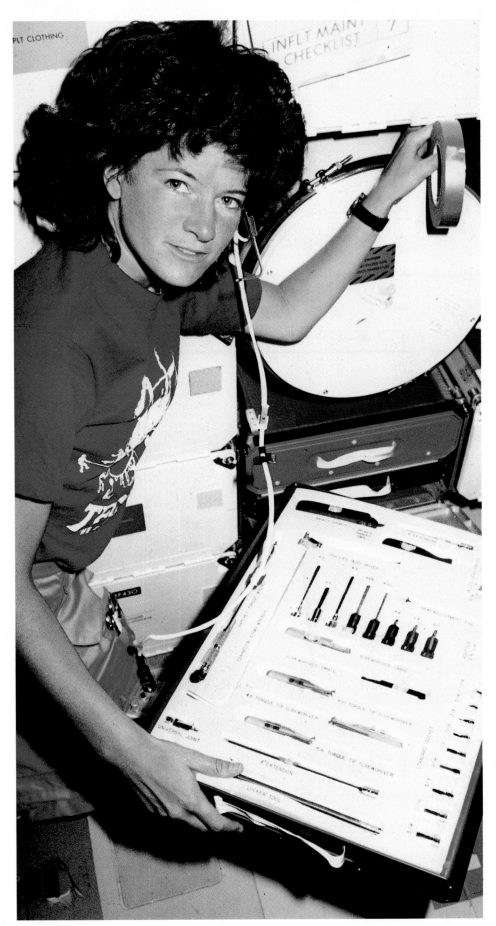

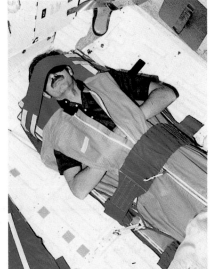

Straps keep Mission Specialist Dale Gardner (above) secure in his berth as he sleeps. Without the straps, he would float about the compartment.

Everything in its place: Sally Ride, a mission specialist, displays a tool box aboard the shuttle. Each tool is enclosed in a form-fitting slot to keep it from floating away. The boxes fit into drawers when not in use. Dr. Ride, a physicist, was the first American woman in space. She took part in a 1983 shuttle mission. As exploration advances, people with many different kinds of specialties will travel into space.

89

NASA (BOTH)

Space mechanic James van Hoften (above) repairs the satellite Solar Max in Challenger's *open cargo bay. Van Hoften, a hydraulic engineer, plants his feet in stirruplike restraints to stay in place. Solar Max, launched to monitor the sun's activities, had blown some fuses. It couldn't keep properly turned toward the sun.* Challenger's *crew maneuvered the shuttle alongside the satellite. Then, using a long robotic arm, they cradled Solar Max in the cargo bay for the needed repairs.*

Mission completed, van Hoften retreats along the cargo bay. Solar Max rests in the repair dock behind him. The robotic arm, left, that helped grasp and position the satellite will place it back into orbit. As more and more satellites are carried into space, repair work like this will become routine.

Jacks-of-all-trades, astronauts put together sections of a permanent manned space station (left). Plans are now being drawn up for the project. A shuttle will ferry the parts from earth. This painting shows the station as construction begins.

The completed station (right) will serve as a launching site for space missions, as a laboratory, and as a specialized factory. It will orbit about 250 miles (402 km) from earth. Astronauts will live aboard for up to six months at a time. Solar panels will collect the energy needed to run the station.

(Continued from page 88) mid-deck a bit too hard and finding myself sprawled on the opposite wall. But I mastered zero g quickly. Soon I felt graceful and could fully control my body and motion."

Eating in space presents some difficulties. Dry foods can easily scatter because of weightlessness. Most space foods are moist or prepared in thick sauces. Such foods cling better to spoons and forks.

You'll also have to get used to a new daily schedule. The shuttle hurtles around the earth at 17,500 miles an hour (28,163 km/h). The sun appears to rise or set every 45 minutes.

If traveling on the shuttle isn't for you, you still can participate in the space program. In various programs run by NASA, individuals and groups send up approved experiments. Young people taking part in these programs have studied such things as the effect of zero gravity on insect behavior and on plant growth. NASA officials say such experiments could provide as

much valuable information as more costly projects.

In just a few years, the shuttle will begin perhaps its most exciting mission. It will start hauling parts for a permanently orbiting manned space station. "We can follow our dreams to distant stars," said President Ronald Reagan in announcing plans for the space station in January 1984. The project, he continued, will contribute to "peaceful economic and scientific gains."

A permanently orbiting space station will enable scientists to live and work in space for long periods. Spacecraft will be able to dock at the station for refueling. The station will serve as the launching site for flights deeper into space.

Several units called modules will make up the station. There will be modules for living and for working. Separate modules will hold supplies. The supply modules will be replaced by shuttle as needed.

"A series of 4 to 6 shuttles will lift the modules into

orbit about 250 miles [402 km] above the earth," says Donna J. Mikov, an official with the Boeing Company, in Seattle, Washington. Boeing is preparing designs for parts of the space station. "Each module is about as big as a large house trailer," Mrs. Mikov continues. "An assembled station would be too bulky to launch from earth. Astronauts will put modules together in space like a giant Lego toy."

Six to 12 crew members will live aboard the space station for as long as 6 months at a time. "We'll make the modules as comfortable as possible," says NASA's O'Donnell. "Sleeping quarters will contain a desk with a reading light, and a small TV. There will be lounges and an exercise room. Astronauts will be relieved by new crew members as they complete their missions. They'll return to earth by shuttle."

In the work module, scientists will make certain products more easily than they can be made on earth. The products include medicines that do not combine well in gravity. Pollution-free space also allows specialists to produce purer crystals for computer chips, and optical *(Continued on page 97)*

Home away from home

Orbiting high above the earth, space-station astronauts will enjoy many of the comforts of home. This painting gives you an idea of what the interior of a living module may look like. An astronaut's day might begin in the lavatory. It will consist of toilet facilities (A), a shower compartment (B), and a dressing room (C). In zero gravity, shower water would normally splatter around the room in droplets. A gentle air flow will make the water stream over the astronaut as it would in gravity. Another air system will draw human waste into storage compartments. The waste will be shuttled to earth for disposal. Equipment in the exercise room (D) will help astronauts keep in shape—and give them a chance to view the changing scenery. In the lounges (E and K), astronauts can socialize, watch videotapes, and relax. The chess set, of course, will be magnetic. While some astronauts are working, others will be getting their sleep in private quarters (F, G, I, and J). Astronauts will take turns standing watch in the command-and-control room (H). Here, an astronaut monitors controls. Outside, a shuttle leaves for earth on a resupply mission. In the galley (L), astronauts will warm and eat prepackaged meals. If they go for space walks, they'll get their pressurized suits from a storage room (M). The astronauts will study stars and other space objects from the wraparound observation room (N) in the station's stern. Space officials expect to complete the first phase of the station in 1992. That will come 500 years after another milestone in exploration: Columbus's sighting of the New World.

A scattering of satellites

So many artificial satellites now orbit the earth that scientists have lost count. Their best estimates put the number of working satellites at around 400. Satellites orbit earth at distances ranging from about 250 miles to 22,300 miles (402–35,887 km). Here are some of the satellites and the jobs they do. Solar Maximum Mission Satellite (**A**) studies the sun. You saw Solar Max on pages 90–91. The Upper Atmosphere Research Satellite (**B**) is scheduled for launch in 1989. It will study how people and natural forces affect the earth's atmosphere. The Earth Radiation Budget Satellite (**C**) measures solar radiation received and given off by earth. Farmers—and those who eat farm products—benefit from Landsat D (**D**), which studies the surface of the earth. National Oceanic and Atmospheric Administration 7 (**E**) tracks weather systems. Satellites of this type have helped save tens of thousands of lives worldwide by warning of storms. Cosmic Background Explorer (**F**), set for launch in 1987, will analyze radiation in space. It will help scientists learn about the early structure of the universe. The International Telecommunications Satellite V (**G**) relays telephone calls and television signals around the world. International Ultraviolet Explorer (**H**) helps scientists study space. It analyzes an invisible form of light—ultraviolet—given off by stars and other bodies. Tracking and Data Relay Satellite (**I**) keeps track of other satellites and relays their data, or information, to earth.

(Continued from page 93) fibers that are super clear.

The crews will also repair space hardware. Billions of dollars' worth of scientific equipment is silently circling the earth. Since the Soviet Union launched the first satellite, Sputnik 1, in 1957, more than a thousand vehicles have gone into orbit. Most of them have fallen back into the atmosphere and burned up.

Experts put the number of working satellites at around 400. Almost two-thirds of this hardware belongs to the United States. Astronauts aboard the space station can lengthen the useful lives of satellites by making regular service calls.

To go on a satellite-repair mission, the astronauts will lift off from the space station in a small utility craft. They'll maneuver the vehicle close to a satellite. They'll either take it on board or repair it in space.

Mission accomplished, they'll return to the station.

Scientists are excited about the space station. They're equally excited about something that will come before it. That's the launching, in 1986, of the Hubble Space Telescope.

"Studying stars from the ground through the hazy atmosphere is like studying birds from the bottom of a lake," says William S. Dempsey, an official with the Perkin-Elmer Corporation, in Danbury, Connecticut. Perkin-Elmer is assembling the optical part of the instrument. "The telescope," says Dempsey, "will let us see clearly into the depths of the universe."

For ages, scientists have asked questions about the universe. How was it formed? When? What unknown wonders does space hold? Does life exist on other planets? The Milky Way, the galaxy in which

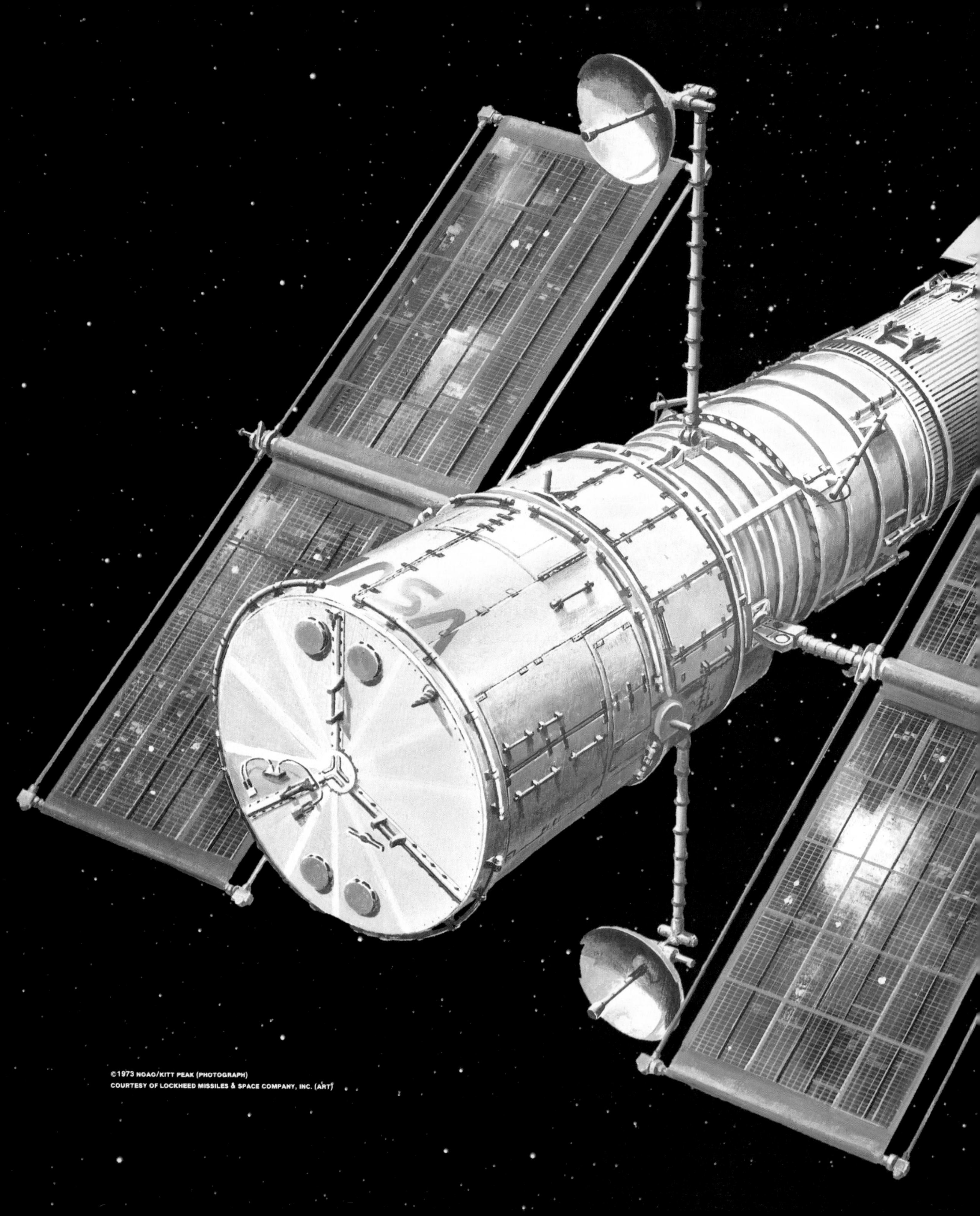
NASA

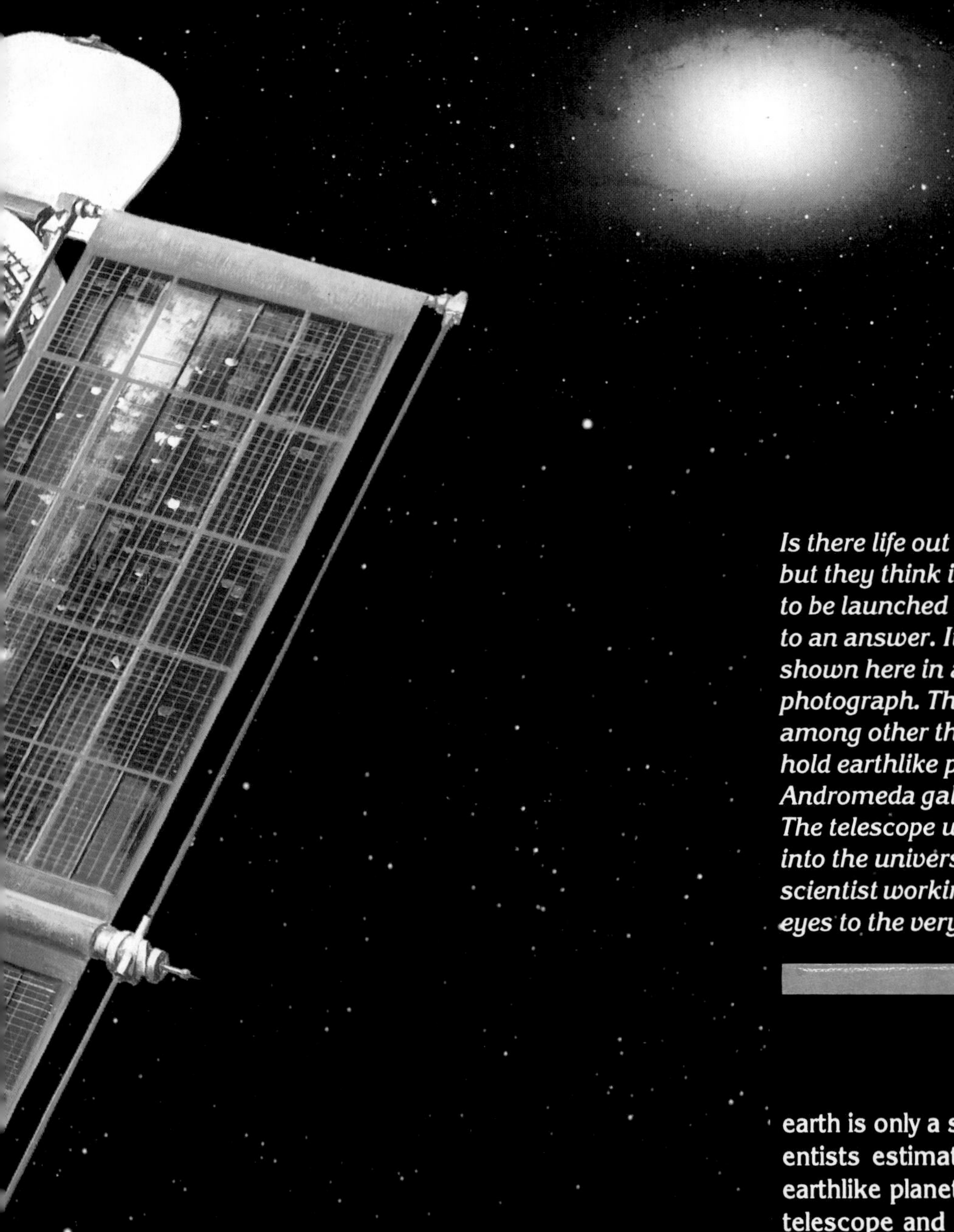

Powerful eyes

Is there life out there? Scientists don't know, but they think it's possible. An instrument soon to be launched may help scientists come closer to an answer. It's the Hubble Space Telescope, shown here in a combination painting and photograph. The telescope will search out, among other things, solar systems that may hold earthlike planets. Here it is viewing the Andromeda galaxy, shown greatly magnified. The telescope will let us see 350 times deeper into the universe than we now can. Says one scientist working on the project: "It will extend our eyes to the very edge of the observable universe."

earth is only a speck, contains trillions of stars. Scientists estimate that there could be 500 *million* earthlike planets in the Milky Way alone. The space telescope and other probes, scientists believe, will find such planets and unlock other age-old secrets.

In reaching for the stars, science is opening whole new worlds. What does that mean to you? At least two things. First, you are certain to benefit from the explosion of scientific knowledge. Second, you could well take an active part in the exploration of deep space. Beyond that, the possibilities are endless—as endless as the universe itself.

CAREERS *in* SCIENCE

Getting a Head Start

By Noel D. Vietmeyer

Science is the most exciting of all explorations. It offers the chance to glimpse things never before seen by humans. It is the key for unlocking the mysteries of the universe.

There are countless explorations yet to make. The more scientists learn, the more mysteries they uncover. No branch of science is complete. Indeed, most sciences are still in their earliest stages.

Consider the breakthroughs described in this book: computers, lasers, robots, gene splicing, bionic body parts, and the rest. Scientists have developed nearly all of them in just the past few dozen years. The future will bring even more dramatic discoveries.

Today's young people will find exciting careers on tomorrow's frontiers of knowledge. Does a career in science appeal to you? To get a head start, you might try doing these things:

• Visit science museums and enter science fairs, such as the one shown on these pages.

• Read all you can about science and technology. Find out what problems scientists are trying to solve.

• Learn how to use a computer. You needn't become a computer whiz—but do learn to make computers work for you.

• Join or start a school science club.

• Meet or read about a leading scientist who is working in a field that interests you. Follow his or her progress.

As a scientist, you will face unlimited challenges . You'll find high adventure. Like a modern-day Columbus, you may discover new worlds.

Dr. Vietmeyer is a scientist with the National Academy of Sciences, in Washington, D. C. You met him in chapter 2. He specializes in discovering new resources for use in poor countries.

A home-built robot draws an audience at the 1984 International Science-Engineering Fair, in Columbus, Ohio. Attending such fairs can give you a firsthand look at the challenges science offers.

Top awards at the fair went to Ann Davis (left, in small picture) and Tim Thrailkill. Ann, who lives in Chicago, Illinois, did heart research. Tim, of Melbourne, Florida, worked with live vaccines.

Index

Bold type indicates illustrations; regular type refers to text.

Additional Reading

Readers may want to check the *National Geographic Index* and the *World Index* in a school or a public library for related articles and to refer to the following books. ("A" indicates a book for readers at the adult level.)

Computers and Lasers: Ardley, Neil, *Computers,* Warwick, 1983. Asimov, Isaac, *How Did We Find Out About Computers?,* Walker, 1984. Berger, Melvin, *Computers in Your Life,* Crowell, 1981. D'Ignazio, Fred, *Small Computers: Exploring Their Technology and Future,* Franklin Watts, 1981. Englebart, Stanley L., *Miracle Chip: The Microelectric Revolution,* Lothrop, Lee & Shephard, 1979. Filson, Brent, *Exploring With Lasers,* Julian Messner, 1984. Jacobsen, Karen, *Computers,* Children's Press, 1982. Jespersen, James, and Jane Fitz-Randolph, *RAMS, ROMS, and Robots: The Inside Story of Computers,* Atheneum, 1984. Lewis, Bruce, *What Is a Laser?,* Dodd, Mead, 1979. National Geographic Society, *Frontiers of Science,* 1982 (A).

Industry: Chester, Michael, *Robots: Facts Behind Fiction,* Macmillan, 1983. D'Ignazio, Fred, *Working Robots,* Elsevier/Nelson, 1982. Ford, Brian J., *Microbe Power,* Stein and Day, 1976 (A). Henson, Hilary, *Robots,* Warwick, 1982. Kiefer, Irene, *Poisoned Land,* Atheneum, 1981. Kleiner, Art, *Robots,* Raintree, 1981. Millard, Reed, *Clean Air–Clean Water for Tomorrow's World,* Julian Messner, 1977. Milton, Joyce, *Here Come the Robots,* Hastings House, 1981. National Geographic Society, *How Things Work,* 1983, and *Small Inventions That Make a Big Difference,* 1984.

Energy and Food: Adams, Florence, *Catch a Sunbeam: A Book of Solar Study,* Harcourt Brace Jovanovich, 1978. Asimov, Isaac, *How Did We Find Out About Solar Power?,* Walker, 1981. Branley, Franklyn M., *Feast or Famine? The Energy Future,* T. Y. Crowell, 1980. Branley, Franklyn M., *Water for the World,* T. Y. Crowell, 1982. Carey, Helen H., *Producing Energy,* Watts, 1984. Davis, George, *Your Career in Energy-Related Occupations,* ARCO, 1980. McDonald, Lucille, *Windmills: An Old-New Energy Source,* Elsevier/Nelson, 1981. Pringle, Laurence, *Nuclear Power: From Physics to Politics,* Macmillan, 1979. Satchwell, John, *Energy at Work,* Lothrop, Lee & Shepard, 1981. Shuttlesworth, Dorothy F. and Shuttlesworth, Gregory J., *Farms for Today and Tomorrow,* Doubleday, 1979. Watson, Jane Werner, *Alternative Energy Sources,* Watts, 1979. Watson, Jane Werner, *Deserts of the World: Future Threat or Promise?,* G. P. Putnam's Sons, 1981.

Transportation: Ardley, Neil, *Transport on Earth,* Franklin Watts, 1981. Georgano, G. N., *Transportation Through the Ages,* J. M. Dent & Sons, 1972 (A). Hellman, Hal, *Transportation in the World of the Future,* M. Evans, 1974.

Medicine: Facklam, Margery and Howard, *From Cell to Clone: The Story of Genetic Engineers,* Harcourt Brace Jovanovich, 1979. National Geographic Society, *Your Wonderful Body!,* 1982, and *Messengers to the Brain: Your Fantastic Five Senses,* 1984. Silverstein, Alvin and Virginia, *The Genetics Explosion,* Four Winds, 1980.

Space: Branley, Franklyn M., *Space Colony: Frontier of the 21st Century,* Elsevier/Nelson, 1982. Fichter, George S., *The Space Shuttle,* Franklin Watts, 1981. National Geographic Society, *The Amazing Universe,* 1975 (A); *Hidden Worlds,* 1981; and *National Geographic Picture Atlas of Our Universe,* 1980. Poynter, Margaret, and Michael J. Klein, *Cosmic Quest: Searching for Intelligent Life Among the Stars,* Atheneum, 1984. Turnill, Reginald, *Space Age,* Frederick Warne, 1980 (A).

ROGER RESSMEYER/STARLIGHT

Cover: Its rockets spitting fire, the space shuttle Challenger *lifts off from its launchpad at Cape Canaveral, Florida. Space missions, such as this one in 1983, help scientists learn more about earth and the endless universe beyond.*

Consultants

Glenn O. Blough, LL.D., University of Maryland—
Educational Consultant
Violet A. Tibbetts—*Reading Consultant*
Phyllis G. Sidorsky, National Cathedral School—*Consulting Librarian*

Nicholas J. Long, Ph.D.—*Consulting Psychologist*

The Special Publications and School Services Division is grateful to the individuals named or quoted within the text and to those cited here for their generous assistance:

Marcia Amsterdam and Susan Stockman, Damon Biotech, Inc.; Joseph A. Bartell and Robert D. Clarke, U. S. Department of Energy; Mike Bergey, Bergey Windpower, Inc.; Carl Blesch, AT&T Bell Laboratories; Richard Burton, The Singer Company; Harlen Chapman, ARCO Solar, Inc.; Charles M. Citrin, M.D., The Neurology Center; T. Rao Coca, IBM Corporation; Mark Cocroft and Anthony V. Gagliardi, General Motors Corporation; James C. Elliott, Goddard Space Flight Center (NASA).

Anna Ganahl, Art Center College of Design; John Gladys, Duke University; Thomas Gray, American Wind Energy Association; Kindred Hawes, Southern California Edison, Inc.; Mark Hess, Kennedy Space Flight Center (NASA); Neal G. Hicks, Environmental Research Laboratory; Tom Hiester, Flowind Corporation.

Jay Holmes, U. S. Department of Energy; Stephen Jacobsen, M.D., Richard Johnson, M.D., David Knutti, M.D., and Harold Sears, M.D., University of Utah; Ruth Kitchen, New Rochelle Hospital Medical Center; Jeff Levinsky and Joan Targ, Stanford University; Sandra Lo Pinto, Flow Systems, Inc.

John C. Mazziotta, M.D., UCLA School of Medicine; Gary McAlister, Maxwell Laboratories; Sam McGee, Laser Images, Inc.; William J. McNeil, Oregon Aqua Foods; John McVeigh, Federal Communications Commission; W. Thomas Neal, General Electric Corporation.

Gordon M. Pfeiffer, Chrysler Canada, Ltd.; Charles Redmond, NASA Headquarters; James Reggia, M.D., University of Maryland; Henry Robitaille, Walt Disney World Epcot Center; Howard A. Rusk, M.D., New York University Medical Center.

Alexander Schriener, Jr., Union Oil Geothermal; Dorothy Schriver, Science Service; Sharon Wanglin, Hubble Space Telescope Science Institute; Kurt Wheeler, Sungene Technologies Corporation; William L. Withuhn, Smithsonian Institution; Bill Wolverton, National Space Technology Laboratories (NASA); Timothy M. Wright, Hospital for Special Surgery.

Composition for SCIENCE: IT'S CHANGING YOUR WORLD by National Geographic's Photographic Services, Carl M. Shrader, Director; Lawrence F. Ludwig, Assistant Director. Printed and bound by Holladay-Tyler Printing Corp., Rockville, Md. Film preparation by Catharine Cooke Studio, Inc., New York, N.Y. Color separations by the Lanman-Progressive Co., Washington, D. C.; and NEC, Inc., Nashville, Tenn. FAR-OUT FUN! printed by McCollum Press, Inc., Rockville, Md.; Classroom Activities folder produced by Mazer Corp., Dayton, Ohio.

Library of Congress Cataloging in Publication Data
Martin, Paul D., 1946–
    Science: it's changing your world.
    (Books for world explorers)
    Bibliography: p.
    Includes index.
    SUMMARY: Discusses scientific breakthroughs and future projections in industry, transportation, medicine, space, and other areas.
    1. Science—Juvenile literature. 2. Technology—Juvenile literature. [1. Science. 2. Technology. 3. Forecasting] I. Egger, Antoinette. II. Title. III. Series
Q163.M367   1985   500   85-2936
ISBN 0-87044-516-2 (regular edition)
ISBN 0-87044-521-9 (library edition)

It's Changing Your World

PUBLISHED BY
THE NATIONAL GEOGRAPHIC SOCIETY
WASHINGTON, D. C.

Gilbert M. Grosvenor, *President*
Melvin M. Payne, *Chairman of the Board*
Owen R. Anderson, *Executive Vice President*
Robert L. Breeden, *Vice President,
Publications and Educational Media*

PREPARED BY THE SPECIAL PUBLICATIONS
AND SCHOOL SERVICES DIVISION

Donald J. Crump, *Director*
Philip B. Silcott, *Associate Director*
William L. Allen, *Assistant Director*

BOOKS FOR WORLD EXPLORERS

Pat Robbins, *Editor*
Ralph Gray, *Editor Emeritus*
Ursula Perrin Vosseler, *Art Director*

STAFF FOR *SCIENCE: IT'S CHANGING YOUR WORLD*
Ross Bankson, *Managing Editor*
Charles E. Herron, *Picture Editor*
Viviane Y. Silverman, *Assistant Art
Director/Designer*
Paul D. Martin, *Writer (text)*
Antoinette Egger, *Writer (legends)*
Donna B. Kerfoot, M. Linda Lee,
Tee Loftin, *Researchers*
Nancy J. Harvey, *Editorial Assistant*
Bernadette L. Grigonis, *Illustrations Assistant*
Janet A. Dustin, *Art Secretary*

STAFF FOR *FAR-OUT FUN!*: Patricia N. Holland,
Project Editor; Martha C. Christian, *Text Editor;*
Roz Schanzer, *Artist*

ENGRAVING, PRINTING, AND PRODUCT MANUFACTURE:
Robert W. Messer, *Manager;* George V. White,
Production Manager; George J. Zeller, Jr., *Production
Project Manager;* Mark R. Dunlevy, David V. Showers,
Gregory Storer, *Assistant Production Managers;* Mary
A. Bennett, *Production Assistant;* Julia F. Warner,
Production Staff Assistant

STAFF ASSISTANTS: Elizabeth A. Brazerol, Dianne T.
Craven, Carol R. Curtis, Lori E. Davie, Mary Elizabeth
Davis, Ann Di Fiore, Eva A. Dillon, Annie Hampford,
Virginia W. Hannasch, Joan Hurst, Artemis S.
Lampathakis, Katherine R. Leitch, Cleo Petroff, Pamela
Black Townsend, Virginia A. Williams, Eric W. Wilson

MARKET RESEARCH: Mark W. Brown, Joseph S. Fowler,
Carrla L. Holmes, Meg McElligott Kieffer, Barbara
Steinwurtzel, Judy Turnbull

INDEX: George I. Burneston III